Computational Intelligence Methods and Applications

The books in this series explain methods developed in computational intelligence (including evolutionary computing, neural networks, and fuzzy systems), soft computing, statistics, and artificial intelligence, and their applications in domains such as heuristics and optimization; bioinformatics, computational biology, and biomedical engineering; image and signal processing, VLSI, and embedded system design; network design; process engineering; social networking; and data mining.

This series includes monographs, introductory and advanced textbooks, contributed volumes and state-of-the-art collections. Furthermore, it supports Open Access publication mode.

Bassem Jarboui • Said Toumi • Patrick Siarry
Editors

Circular Economy Supply Chains

Optimizing via Data Science

Springer

Editors
Bassem Jarboui
Hamdan Bin Mohammed Smart University
Dubai, United Arab Emirates

Said Toumi
College of Business Administration
Majmaah University
Al-Majmaah, Saudi Arabia

Patrick Siarry
LiSSi Lab
Paris-Est Créteil University
Paris, France

ISSN 2510-1765 ISSN 2510-1773 (electronic)
Computational Intelligence Methods and Applications
ISBN 978-981-96-7904-1 ISBN 978-981-96-7905-8 (eBook)
https://doi.org/10.1007/978-981-96-7905-8

This Springer imprint is published by the registered company Springer Nature Singapore Pte Ltd.
The registered company address is: 152 Beach Road, #21-01/04 Gateway East, Singapore 189721, Singapore

Preface

The **Circular Economy** aims to create closed-loop systems where resources are used efficiently, products are designed for longevity and recyclability, and waste is minimized through innovative strategies such as recycling, reuse, and remanufacturing. **Circular supply chain**, as one of the important configurations of the **circular economy**, has received considerable attention in sustainability matters. Within the framework of the **circular supply chain** several key objectives emerge, each addressing different aspects of resource management, waste reduction, and environmental impact mitigation. This book aims to detail the application of **operations research** and **data science** methods as decision support systems to manage complex constraints. It presents the most recent advancements and real-world applications of these methods within the framework of **supply chain management** in the **circular economy**. Additionally, it explores case studies that highlight successful implementations and the benefits achieved. The book offers comprehensive insights and practical applications, serving as a valuable guide for students, researchers, and professionals looking to drive sustainability in their organizations and communities. Divided into two sections, it explores innovative approaches to enhance circularity and sustainability in supply chain management.

Section 1: Optimizing Circularity with Operations Research
This section discusses how different **operations research** methods can be used to address challenges and opportunities in circular supply chains.

Chapter 1, "Leveraging Optimization for the Circular Economy: A Reverse Logistics Routing and Location Perspective" (J. Sarkis and M. Jahanafroozi), provides an overview of reverse logistics focusing on optimization and simulation techniques to enhance efficiency. It explores key optimization methods, such as linear programming and multi-objective optimization, alongside simulation techniques like discrete-event simulation and agent-based modeling. Special attention is given to transportation in Reverse Logistics, highlighting the synergies between optimization and simulation. The chapter also discusses emerging trends, future directions, and technological advances in the circular economy, while identifying gaps in existing literature and suggesting areas for future research.

Chapter 2, "Designing a Closed-Loop Supply Chain Network Considering Uncertain Dynamic Demand and Production Quality" (T. H. Hejazi et al.), presents a closed-loop supply chain network design that integrates forward and reverse logistics, focusing on economic performance, environmental impact, resource utilization, and sustainability. A bi-objective model is developed to minimize costs and environmental impacts, considering imperfect production quality and demand uncertainty, with forecasts generated using the vector autoregressive model. Robust optimization techniques protect the network against uncertainty. Numerical examples and sensitivity analysis validate the model, providing valuable insights for effective CLSC management.

Chapter 3, "Applying a Simulated Annealing Embedded Genetic Algorithm for Solving a Concurrent Stochastic Supply Chain and Product Family Design with the Location-Allocation of Retailers" (T. H. Hejazi et al.), addresses a product family design problem in a stochastic three-echelon supply chain under demand uncertainty, considering low-cost and high-quality product preferences. It includes both mandatory and optional modules for customization, minimizing waste and supporting sustainability. The model optimizes retailer locations to reduce transportation costs and maximize profit. A hybrid genetic algorithm and simulated annealing (GASA) are used to solve the NP-hard problem, with GASA outperforming exact methods.

Chapter 4, "Navigating Circular Horizons: Unveiling the Synergy of Operations Research in Sustainable Supply Chain Management" (H. Chabchoub et al.), reviews how Operations Research techniques like optimization, simulation, and decision analysis improve efficiency in Circular Supply Chain Management. It covers applications in closed-loop supply chains, green logistics, and waste reduction, while addressing challenges like data complexity and sustainability metrics. The chapter also identifies research gaps and opportunities to strengthen the integration of Operations Research with Sustainable Supply Chain Management, offering insights for academics and practitioners.

Chapter 5, "Theory and Research Concerning the Circular Economy Model and Future Trend" (H. R. Yazgan et al.), reviews the Circular Economy approach, analyzing studies from 1994 to 2023 and highlighting emerging trends. It critiques the Linear Economy model's limitations in addressing climate change, biodiversity loss, and resource scarcity, showcasing sustainable alternatives. The study explores the impact of Artificial Intelligence, Industry 4.0, and Operational Research on CE, based on a systematic review of literature from 2007 to 2023. It also proposes integrating Artificial Intelligence with the CE's 9R framework, offering insights for implementing CE practices in production.

Chapter 6, "Sales Center Location, Green Open Vehicle Routing for Product Transport in the Urban Context, and Their Future Perspectives with the Circular Economy" (J. R. Melquiades and L. R. Cerna), optimizes product acquisition and delivery in urban areas to improve service and sustainability. It uses a mixed integer linear programming model and a hybrid algorithm combining k-means clustering and local search. The approach enhances service while reducing fuel consumption and CO_2 emissions. Implemented in GLPK, the model delivers optimal results for small scenarios (up to 19 customers) and for larger scenarios (30–50 customers) the hybrid algorithm obtains good results in short processing times.

Section 2: Data Science for Smarter Circular Supply Chains
This section highlights the importance of **data science** and **data mining** in **circular supply chains**.

Chapter 7, "Implementing Sustainability and Project Management Outcomes" (R. Anvari and F. Moradi), conducts a comprehensive literature review on the intersection of sustainability and project management. Key topics explored include the impact on access to capital, customer loyalty, employee morale and productivity, supply chain improvements, long-term operational performance and efficiency, shareholder value enhancement, risk reduction, and talent recruitment and retention.

Chapter 8, "Driving Sustainable Circularity with Reverse Logistics in Halal Sectors to Degrade Ecological Footprints: Evidence in ASEAN-5" (K. Musari and M. R. P. Sakti), examines the circular economy and reverse logistics in the Halal sectors of the ASEAN-5 to reduce ecological footprints. As leaders in the global Halal industry, the ASEAN-5 can influence other sectors to adopt sustainable practices. Through literature review and netnography, the chapter discusses ecological footprints, the state of Halal sectors, and the adoption of circularity and reverse logistics in these industries. The study finds that while some ASEAN-5 Halal sectors are progressing toward sustainability, reverse logistics is not yet widely implemented.

Chapter 9, "Circular Logistics and Operations: Transportation Across a Product's Lifecycle" (E. Demir et al.), explores the transition to a circular economy, where consumers shift from being product recipients to suppliers of used goods for repair and remanufacturing. It proposes a data-driven decision support system to optimize resource allocation in CE operations. Using Monte Carlo simulations, the study examines pre-inspection, consolidation, and product lifecycles in a real-life case. The findings show that pre-inspection and consolidation at the retailer are beneficial, with the impact depending on warranty duration and customer behavior.

Chapter 10, "Impact of Demand Forecasting on Environmental Performance of Coffee Supply Chain in Vietnam" (A. Bekrar et al.), examines demand forecasting techniques in the Vietnamese coffee supply chain, showing that LSTM models outperform ARIMAX in reducing carbon and water footprints. It underscores the importance of advanced forecasting methods for enhancing environmental performance and highlights critical factors organizations should consider when adopting them. The study provides practical insights into the role of demand forecasting and the challenges organizations may face during implementation.

Chapter 11, "Exploring Global Food Security Index: A Case Study of Indonesia" (A. Nurmandi et al.), analyzes Indonesia's Global Food Security Index (GFSI), focusing on key dimensions like Affordability, Availability, Quality, Safety, and Sustainability. It considers the country's socioeconomic and political context and offers recommendations for improving food security, such as strengthening legal frameworks, investing in infrastructure, and promoting capacity building. These measures aim to protect Indonesia from food crises and enhance overall food security.

These chapters collectively address various aspects of the **circular supply chain**, focusing on different objectives while contributing to the overarching goal of sustainable development and resource stewardship by exploring case studies and practical real-world applications of **data science** and **operations research**.

Dubai, United Arab Emirates
Al-Majmaah, Kingdom of Saudi Arabia
Créteil, France
February 2025

Bassem Jarboui
Said Toumi
Patrick Siarry

Contents

Contributors

Halimah Abdul Manaf UUM College of Law, Government, and International Studies Universiti Utara Malaysia, Kedah, Malaysia

Mourad Abed LAMIH Laboratory, Université Polytechnique Hauts-de-France, Valenciennes, France

Rendra Agusta PT. Sinergi Visi Utama, Yogyakarta, Indonesia

Derya Eren Akyol Department of Industrial Engineering, Faculty of Engineering, Dokuz Eylul University, Izmir, Turkey

Roya Anvari School of Business and Administrative Studies, The University of Georgia, Tbilisi, Georgia

Gulseli Altintaş Department of Industrial Engineering, Faculty of Engineering, Sakarya University, Sakarya, Turkey

Abdelghani Bekrar LAMIH Laboratory, Université Polytechnique Hauts-de-France, Valenciennes, France

Habib Chabchoub Al Ain University, Abu Dhabi, UAE

Lourdes Ramirez Cerna Optimización en logística y combinatoria, Carrera de Ingeniería de Sistemas, Universidad de Lima, Lima, Peru

Parisa Cheshmeh Department of Industrial Engineering, Amirkabir University of Technology, Rasht, Iran

Emrah Demir PARC Institute of Manufacturing, Logistics and Inventory, Logistics and Operations Management Department, Cardiff Business School, Cardiff University, Cardiff, Wales, UK

Thanos E. Goltsos PARC Institute of Manufacturing, Logistics and Inventory, Logistics and Operations Management Department, Cardiff Business School, Cardiff University, Cardiff, Wales, UK

Taha-Hossein Hejazi Department of Industrial Engineering, Amirkabir University of Technology, Rasht, Iran

Mozhgan Jahanafroozi LAMIH, Université Polytechnique Hauts-de-France, Valenciennes, France

Fatemeh Jahangir Department of Industrial Engineering, Sadjad University of Technology, Mashhad, Iran

Danang Kurniawan PT. Sinergi Visi Utama, , Yogyakarta, Indonesia

Arij Lahmar Dubai Business School, University of Dubai, Dubai, UAE

Thi-Muoi Le CRISS Laboratory, Université Polytechnique Hauts-de-France, Valenciennes, France

Farideh Moradi Beykoz University, , Istanbul, Turkey

Khairunnisa Musari Kiai Haji Achmad Siddiq State Islamic University (UIN KHAS), Jember, Indonesia

Dyah Mutiarin Department of Government Affairs and Administration, University of Muhammadiyah Yogyakarta, Yogyakarta, Indonesia

Thi Thuy Hanh Nguyen University of Economics and Law and Vietnam National University, Ho Chi Minh City, Vietnam

Ebrahim Rezaee Nik Department of Industrial Engineering, Sadjad University of Technology, Mashhad, Iran

Achmad Nurmandi Department of Government Affairs and Administration, University of Muhammadiyah Yogyakarta, Yogyakarta, Indonesia

Andi Luhur Prianto Department of Government Studies, Universitas Muhammadiyah Makassar, Makassar, Indonesia

Muhammad Rizky Prima Sakti University College of Bahrain, Manama, Kingdom of Bahrain

José Rodríguez-Melquiades Optimización en logística y combinatoria, Departamento de Informática, Universidad Nacional de Trujillo, Trujillo, Peru

Joseph Sarkis Business School, Worcester Polytechnic Institute, Worcester, MA, USA

Mohsen Shidpour Department of Industrial Engineering and Management Systems, Amirkabir University of Technology, Tehran, Iran

Areej Aftaab Siddiqui Dubai Business School, University of Dubai, Dubai, UAE

Triwiyanti Triwiyanti PT. Sinergi Visi Utama, Yogyakarta, Indonesia

Sean White Global Head of Repair Services, DSV; PARC Institute of Manufacturing, Logistics and Inventory, Cardiff Business School, Cardiff University, Cardiff, Wales, UK

Shafira Mourina Yasmine PT. Sinergi Visi Utama, Yogyakarta, Indonesia

Harun Resit Yazgan Engineering Faculty, Industrial Engineering, Sakarya University, Serdivan, Turkey

Muhammad Younus Department of Government Affairs and Administration, University of Muhammadiyah Yogyakarta, Yogyakarta, Indonesia
Department of Product Research and Software Development, TPL Logistics Pvt Ltd, Karachi, Pakistan

Sepideh Asadi Zeidabadi Department of Industrial Engineering and Management Systems, Amirkabir University of Technology, Tehran, Iran

Acronyms

AI	Artificial Intelligence
ARIMAX	Autoregressive Integrated Moving Average model
C	Customer
CE	Circular Economy
CL	Circular Logistics
CLSCND	Closed-Loop Supply Chain Network
CSC	Circular Supply Chain
CSCM	Circular Supply Chain Management
EoL	End-of-Life
GASA	Genetic Algorithm and Simulated Annealing
GDP	Gross Domestic Product
GIE	Global Islamic Economy
GMTI	Global Muslim Travel Index
GPS	Global positioning System
ICT	Information and Communication Technology
IoT	Internet of Things
KPI	Key Performance Indicators
LE	Linear Economy
LP	Linear Programming
LSTM	Long Short-Term Memory model
MILP	Mixed Integer Linear Programming
MINLP	Mixed-Integer Non-Linear Programming
ML	Machine Learning
MOMILP	Multi-Objective Mixed-Integer Linear Programming
MOMINLP	Multi-Objective Mixed-Integer Non-Linear Programming
MOO	Multi-Objective Optimization
NLP	Non-Linear Programming
OBS	Optimization-Based Simulation approach
OR	Operations Research
OR	Operational Research
PLC	Product Lifecycle

REC	Recycler
REM	Remanufacturer
RET	Retailer
Rl	Reverse Logistics
SCM	Supply Chain Management
SD	System Dynamics simulations
SDG	Sustainable Development Goals
SDGI	Sustainable Development Goals Index
SSC	Sustainable Supply Chains
SSCM	Sustainable Supply Chain Management
TS	Time Series
UVA	Use Value Appraisal
VAR	Vector Autoregression
VBI	Value-Based Intermediation

Part I
Optimizing Circularity with Operations Research

Chapter 1
Leveraging Optimization for the Circular Economy: A Reverse Logistics Routing and Location Perspective

Joseph Sarkis and Mozhgan Jahanafroozi

Abstract A brief review of reverse logistics literature is presented in this chapter, with an emphasis on optimizing and simulating approaches to enhance efficiency and effectiveness. Economic, environmental, and legal factors drive reverse logistics, which involves the return of goods from their final destination to their origin for value recovery. Various optimization techniques, including linear programming, integer programming, and multi-objective optimization, can be used to address reverse logistics challenges. Discrete-event simulation, system dynamics, and agent-based modeling also provide insights into reverse logistics complexities. These two sets of analytical and modeling tools optimization and simulation in reverse logistics are overviewed in this chapter. We also shift the focus to a more specific and understudied area of RL modeling and transportation. Synergies are highlighted between the approaches, illustrating how their integration, especially for transportation in reverse logistics, enhances the decision-making process. Additionally, the chapter covers emerging trends as well as future directions, focusing on technological advances and sustainability within the broader circular economy context. It concludes with identifying gaps in the existing literature and suggesting potential areas for future study.

Keywords Reverse logistics · Transportation · Circular economy · Optimization · Location-routing problem

J. Sarkis
Business School, Worcester Polytechnic Institute, Worcester, MA, USA
e-mail: jsarkis@wpi.edu

M. Jahanafroozi (✉)
LAMIH, Université Polytechnique Hauts-de-France, Valenciennes, France
e-mail: mozhgan.jahanafroozi@uphf.fr

B. Jarboui et al. (eds.), *Circular Economy Supply Chains*,
Computational Intelligence Methods and Applications,
https://doi.org/10.1007/978-981-96-7905-8_1

1.1 Introduction

Businesses depend heavily on supply chain management (SCM), which comprises five key functions: purchasing, operations, resource management, information workflows, and logistics [18]. Logistics moves goods and services between locations to ensure timely delivery, proper condition, and accurate location. The traditional forward and internal logistics management for organizations is supplemented by external and closing-the-loop logistics often referred to as reverse logistics.

Reverse logistics plays an important role for managing in a circular economy. The scarcity of raw materials, environmental pollution, social responsibility, regulations, and market shifts caused greater interest in reverse logistics (RL) [1]. RL manages the flow of materials and goods from the point of consumption back to the origin, aiming to recapture value or properly dispose of them (Going Backwards: Reverse Logistics Trends and Practices, [s.d.]). Incorporating reverse logistics into traditional supply chains helps businesses manage capital more efficiently, reduce waste, and generate income by recovering products and materials [31]. Businesses are under constant pressure to innovate, differentiate, and operate more efficiently in a highly competitive environment. Resource scarcity, climate change, increased waste generation and management, and contaminated waste risks make companies vulnerable due to their unsustainable operations. For over 200 years, businesses have widely used a linear model of "Take, Make, and Waste" where manufacturers produce goods, sell them to consumers, and the products are discarded after use [47]. These factors have driven a shift toward a new generative model. According to the Ellen MacArthur Foundation, the circular economy is designed to be restorative, using renewable energy, minimizing toxic chemicals, and eliminating waste through careful planning [32]. The circular economy can foster a more sustainable economic system if appropriately planned and maintained. This is where RL plays an important role.

The transition to a circular economy has clarified RL as an approach to handling end-of-life products. RL chains will need to be optimized to enhance circular design and reduce material loss [47]. RL systems are still in their relative infancy because organizations have not viewed these activities as value-adding. In addition, managing RL is complex, involving multiple activities such as collection, inspection, storage, disassembly, and handling outbound materials [34]. Sustainable reverse logistics faces a multiplicity of challenges. High costs, uncertainty in product volume and quality, lack of real-time data, and poor coordination among stakeholders are only some of the most obvious [28, 42, 53, 55].

Optimization and simulation techniques are increasingly used to improve reverse logistics by aiding decision-making in resource allocation, route planning, and inventory management. Optimization reduces costs and boosts efficiency, while simulation allows businesses to model processes, predict outcomes, and test strategies without risk. Industry 4.0 technologies, like the Internet of Things (IoT), artificial intelligence (AI), and big data analytics, enhance optimization and simulation in RL by enabling real-time data, improving decision-making, and increasing process transparency. These advancements help businesses manage reverse logistics more efficiently and sustainably.

This chapter reviews current research and practical optimization applications, and simulation, which are part of Industry 4.0 technologies, in RL a central element of the circular economy. In this thematic review, we explore how these methods address challenges, offering insights into their role in creating more sustainable and resilient supply chains, and highlighting their potential to enhance circular economy practices for environmental and economic benefits.

1.2 Literature Review

The integration of reverse logistics into supply chain management, especially within the circular economy, has gained significant attention in research. This review focuses on the current studies related to optimization techniques, simulation methods, and the role of Industry 4.0 technologies in improving reverse logistics processes.

1.2.1 The Role of Reverse Logistics in the Circular Economy

Reverse logistics in the circular economy brings back materials into the production system to include returning, recycling, remanufacturing, and reusing products and materials to minimize waste and recover resources [39]. For instance, it creates a closed-loop system (closed-loop supply chain), for managing end-of-life and extended producer responsibility efforts such as waste electrical and electronic equipment (WEEE)[39]. RL is the systematic management of raw materials, finished products, inventory in progress, and related data through the movement of raw materials, finished products, and work-in-progress inventory from the destination back to the source with a view to recovering value [39]. RL definitions sometimes include the natural environment in their definition. This inclusion implies belief that RL is environmentally friendly. But, sometimes transportation concerns arise that may make the environmental benefits appear. We touch upon various transportation concerns below in discussing RL activities and processes in this section.

Items typically returned to return operations fall into four categories: reusable items, repair services, remanufacturing, and recycling [19]. This breakdown and characteristic has resulted in an increased concern of RL that is not limited to mere returns management but also emphasizes sustainability, cost reduction, and value recoveryClick or tap here to enter text. RL has gained prominence due to concerns related to environmental concerns, legal and regulatory requirements, and economic considerations. Similarly to forward supply chain logistics, RL is likely to involve multiple organizations involved in a variety of activities [7]. It comprises activities such as collection, separation and inspection, storage, disassembly, shredding, and grinding, and outbound logistics based on the paper Meade et al. [34]. Each of these activities must be completed in accordance with the product type [7]. It is critical to note that electronic products are processed very differently from textiles, for instance. In addition, the mode of transportation for each of these products may differ. Shipments of textiles may be transported by truck, whereas those of electronic components may be transported by plane. Following this step, it is necessary to determine whether to repurpose, repair, remanufacture, or recycle in order to recoup value or dispose of the item [4]. For instance, if an item is obsolete, it is typically cheaper to recycle than to repair it.

Reverse logistics starts with collection. Collection includes retrieval of goods from consumers, retailers, or other points along the supply chain. A variety of factors drive materials into reverse logistics (RL) channels, including adherence to regulations like WEEE policies and the pursuit of business goals like the retrieval of valuable resources such as lithium [54]. The capture of such rare and not-so-rare materials means that extraction, delivery across the globe, and international regulatory complexities cause fewer environmental and business concerns. These are fundamental advantages of CE and cannot occur without RL. Collection varies with original equipment manufacturers, collection sites, retail outlets, and independent collectors, each having a different reason and approach in the RL network [7]. RL requires careful consideration when developing networks such as selecting collection centers and recovery facilities as part of the design process [43]. Collection and transportation pose significant environmental concerns with substantial carbon and pollutant emissions that would need to be included in any model. Even activities between two organizations can include collection and back-haul transportation, while providing reusable transportation packages [10].

After collection, inspection and sorting in RL are typically completed. The sources of RL material flows vary-there also may be considerable differences in quality and condition of materials. Care should be taken to determine where and when these materials need to be directed-for example, and it is also possible to store returned materials directly for reuse [7]. Returned items can be sorted into a broad set of categories from defective, damaged, to unused before being sent to RL facilities. Transportation issues arise with different modes of transportation selected or assigned depending on product and material characteristics. For example, low-value, high-density, and weight waste, such as construction materials, may be shipped by boat, while high-value waste electronics, such as used cell phones, may be shipped by airplane. Thus, the mode of transport and selection may play important roles.

Storage for RL inventory and flows is another stage in RL [7]. Once again, depending on the material and quality type, decisions on where and how to store them become a nontrivial issue. By carefully storing returned products, RL systems can strategically balance supply and demand, ensuring that resources are readily available for necessary processes like refurbishment, recycling, or resale. The storage and facility system for warehousing returned and end-of-life materials requires various transportation planning such as moving materials from collection to storage locations and internal organizational transportation and logistics (materials management) requirements. Planning these within a broader network requires significant analytical considerations, some of which will be covered in latter sections in this chapter.

Product or material characteristics also relate to the next stage of disassembly operations, which include activities such as shredding and grinding. These activities can occur within one facility or across several facilities [7]. The planning of these operations will require determination of transportation feasibility, timing, and selection—to name a few decisions and characteristics of transportation with RL that would need to be considered. We return to this issue from a broader vehicle-routing and transportation planning perspective, later. When a product contains multiple components, disassembly operations can be distributed among several specialized facilities. Each facility can handle a particular component of the recycling or refurbishment process. For example, Metech, an electronics asset recovery company, has the capability to shred materials but cannot recover certain elements and needs to send them thousands of miles for gold and precious metal recovery [40]. RL partners can and do collaborate to help ensure materials are handled efficiently and in environmentally responsible ways. RL requires tracking flow of materials throughout their lifecycle, from disassembly to redistribution and recycling. This tracking and planning includes various transportation requirements within each of these processes. An activity within and at the end of RL activities is outbound logistics. This activity refers to especially to transportation of finished goods (after all RL and circular economy processing has been completed) from a company's location to its end users—who may happen to be part of the forward supply chain and traditional manufacturers and fabricators.

It is important to recognize that outbound logistics and transportation can exhibit considerable diversity, resembling the wide range of requirements seen in other aspects of RL. The outbound logistics process can involve a variety of activities such as transportation, storage, packaging, and delivery—each of which would need to be incorporated into optimization models to holistically model the RL environment. It can also involve different modes of transport, such as air, rail, and road, as well as different types of packaging, such as boxes and pallets. Outbound logistics processes in RL can also vary depending on the type of product, such as perishable or nonperishable goods.

Thus although RL can improve sustainability, enhance recovery rates, reduce CO2 emissions, and require collaboration among stakeholders to meet regulations [52], there are significant transportation planning and design issues that can be managed through simulation and optimization tools, as well as broader inclusive technologies such as Industry 4.0. We now present a general overview of RL optimization background, before we specifically target transportation concerns.

1.2.2 An Overview of Reverse Logistics Optimization Techniques and Applications

Quantitative and analytical models for RL have existed since before the term sustainability became popular. In one of the more popular reviews, [10] provides a simplistic model of actors including collectors and recyclers but even during that period a variety of models and issues were presented. The quantitative modeling considered inventory control in systems with return flows which was a major approach. These inventory models reviewed included deterministic, stochastic models with repair and product recovery models. A second category of models—some of these issues defined in the RL sections—related to the production planning and reusing parts and materials, selection of recovery options, and scheduling of recovering environments as part of operational optimization and simulation. Part of these internal activities included the design and inclusion of materials planning and reverse bills of material, for example. This was the scope of research, albeit very early on, of modeling efforts during that early period. Specific external relationships and transportation concerns were not as evidenced in modeling during these earlier periods. In fact, even later reviews of RL and quantitative analyses, including simulation and optimization, saw few models explicitly considering transportation.

There are some examples, tools involved in RL that have been extensively studied in terms of optimizing performance. Various methodologies including mixed-integer linear programming, stochastic mixed-integer programming, mixed-integer nonlinear programming, simulation studies, agent-based modeling, system dynamics, and geolocation procedures grown over the past three decades. We will not cover model types and characterizations in this chapter. However, recent reviews show an expansion of where and the type of applications for "closing-the-loop" and circular practices, many of which include RL. For example, Table 1.1 (adapted from Mahmoum Gonbadi et al., [33]) shows specific optimization tools (simulations were missing in their review) that summarized papers from a structured literature review. They considered quantitative closed-loop decision models and applications and further categorized them into strategic, tactical, and operational levels of decision-making. Many of these decisions are related to transportation modeling, but most consider multiple dimensions. According to them, most models focused on strategic concerns.

Although transportation was included in some of the contexts, it was not clear how many delved deeply into transportation modeling for RL. Few if any, covered transportation within the context of technological enablers although interest has been growing in this area according to the paper Mahmoum Gonbadi et al. [33]. As can be seen, by the categorization of decision type, the decisions seem to be "technology agnostic."

Table 1.1 Organizational decision and optimization modeling levels for closed-loop supply chains (Source: Adapted from Mahmoum Gonbadi et al. [33])

Decision level	Decision type
Strategic	Number of facilities
	Facility location
	Facility capacity
	Facility scale
	Technology type
	Transportation channels
	Product design
	Transportation mode
Tactical	Allocations
	Supplier selection
	Inventory levels
	Pricing decisions
	Discount level
	Transportation amount
	Planning activities
	Vehicle selection
	Fleet composition
Mixed strategic tactical	Facility location allocation
	Facility location inventory management
	Facility location product flow
	Number of facilities supplier selection
	Transportation mode transportation quantity
Strategic operational	Facility location transportation scheduling
Tactical operational	Transportation quantity reorder point
All levels	Facility location inventory level vehicle-routing planning
	Facility location allocations flowshop scheduling
	Facility location allocations vehicle-routing planning
	Facility location transportation and inventory decisions production planning

1.3 Optimization Models for Transportation Routing and Location for Reverse Logistics

In this section, we provide some thematic examples of various optimization modeling approaches and their applications to vehicle routing, transportation, and location analysis and RL considerations. Some of these decisions and models are included in the broad variety of supply chain decisions in the previous section. We introduce some characteristics of the various themes and models. We do not exhaustively review the models and applications but provide some insights into the complexity of decisions and concerns, especially given transportation and vehicle-routing activities. We also return to our overview of RL activities including collection, sorting and inspection, storage, disassembly and processing, and outbound logistics, providing a linkage, or lack thereof, by these activities to current studies. The focus is on optimization models in each of these RL activities and specifically mathematical programming approaches.

A limited number of studies have been focused on location-routing problems in RL, despite its importance at all levels of decision-making. Location-routing problems are crucial because they combine the challenges of determining optimal facility locations with the complexities of route planning. This integration helps to minimize costs and improve efficiency in logistics and supply chain management. Effective solutions can lead to significant savings and enhanced service levels. Thus, we tend to specifically focus on the location-routing problem.

1.3.1 An Overview of the Published Literature

Table 1.2 presents a summary of various studies on RL and related fields. Aspects include different modeling approaches, solution techniques, objectives, and the consideration of certainty or uncertainty in parameters. Each of these categories is now defined.

Year and Reference: Each row corresponds to a specific study, starting from the most recent (2024) to the oldest (2006). Model type: Indicates the mathematical or computational model used in each study. Models include MILP (mixed-integer linear programming), MIP (mixed-integer programming), LRP (location-routing problem), and multi-objective optimization. There are specific aspects of these models that are considered such as bi-objective optimization, and variations on MILP models, such as two-stage and flow models.

Solution: Describes the type of algorithm or method used to solve the model. These include heuristic methods, exact algorithms, metaheuristic approaches, and hybrid algorithms. Many of the models were able to be solved without special algorithms and used standard software as CPLEX. Yet, from a contribution perspective, some authors took advantage of the characteristics of the models to solve with heuristics or metaheuristics. A very common approach was a mixed methodology hybrid approach. Hybrid approaches combine multiple techniques for enhanced performance and solving complex logistics problems. Techniques like genetic algorithms, tabu search, and volleyball premier league algorithms are used for more complex and large-scale problems, for complex MILP-like formulations.

Objective functions: Economic, environmental, or social goals each of which is typically associated with RL and its activities were used as general goals to optimize. Common objectives include minimizing total costs, maximizing profit, minimizing transportation costs, or maximizing customer satisfaction. These most common objectives clearly focused on business and economic goals, which included improving efficiency. Environment: Objectives such as minimizing carbon emissions and environmental impacts are often secondary to economic factors. But, these

Table 1.2 Summary listing of reverse logistics, routing, and location literature used in this study analysis

Reference	Year	Model type	Solution	Objective functions	Certain/uncertainty	Application area
Hosseini Dehshiri & Amiri [16]	2024	MILP	Exact	**Economic:** Cost minimization, **Environmental:** Carbon reduction	**Uncertain**	Closed-loop supply chains
Lei et al. [22]	2024	MILP	Exact/heuristic	**Economic:** Minimize total cost	**Uncertain**	Reverse logistics (collection-disassembly-delivery)
Luo et al. [30]	2024	MILP	Exact/heuristic	**Economic:** Minimize operational costs, **Social:** Maximize service quality	**Uncertain**	Waste collection and transportation
S. Sun et al. [50]	2023	MILP	Metaheuristic (HVPL)	**Economic:** Minimize total cost	**Uncertain**	Reverse logistics
Zhou et al. [61]	2023	MILP	MOHGATS (Heuristic)	**Economic:** Maximize profit, **Environmental:** Carbon reduction	**Uncertain**	Reverse logistics for eco-packages
Q. Liu et al. [26]	2023	MILP	Heuristic	**Economic:** Minimize total cost	**Certain**	Underground waste collection system
Han et al. [15]	2023	MILP	Heuristic	**Economic:** Shortest delivery time, Lowest transportation cost	**Uncertain**	Healthcare logistics and distribution
Oliveira Neto et al. [39]	2023	MILP	Exact/heuristic	**Economic:** Maximize profit, Minimize total cost	**Uncertain**	Reverse logistics for waste electrical and electronic equipment
Huang et al. [17]	2023	Two-stage robust MILP	Exact/heuristic	**Economic:** Minimize total cost under uncertain demand	**Uncertain**	Reverse logistics for bike-sharing systems
Govindan et al. [13]	2022	Bi-objective MILP	Augmented epsilon-constraint method	**Economic:** Minimize total costs	**Uncertain**	Medical waste management and green reverse logistics
Y. Zhao et al. [60]	2021	Flow-based MILP	Hyper-heuristic approach	**Economic:** Minimize total costs	**Uncertain**	Reverse logistics

(continued)

Table 1.2 (continued)

Reference	Year	Model type	Solution	Objective functions	Certain/uncertainty	Application area
J. Li et al. [23]	2021	Heuristic	Heuristic	**Economic:** Minimize transport cost, **Social:** Maximize customer satisfaction	**Uncertain**	E-commerce logistics
Rabbani et al. [44]	2020	Bi-objective MIP	CPLEX	**Economic:** Minimize cost, **Social:** Minimize population risk	**Uncertain**	Industrial waste management
Pingping and Yu [41]	2020	Bi-level model	Metaheuristic	Not mentioned	**Uncertain**	Reverse logistics
Zandkarimkhani et al. [58]	2020	Bi-objective MILP	Exact	**Economic:** Minimize cost, Social: Maximize site value	**Uncertain**	Sustainable open-loop supply chain network
Govindan et al. [12]	2020	Two-stage MILP	Exact	**Economic:** Minimize costs, shortages	**Uncertain**	Circular supplier selection, closed-loop supply chain
Bagheri Hosseini et al. [2]	2019	MIP	Heuristic	**Economic:** Minimize total cost	**Certain**	Reverse logistics, remanufacturing, waste management
Nadizadeh and Kafash [36]	2019	MILP	Heuristic	**Economic:** Minimize fixed depot and travel costs	Uncertain	Transportation and logistics
Nowakowski and Mrówczyńska [38]	2018	Artificial immune system	Heuristic	**Economic:** Minimize route length, Optimize container placement	**Uncertain**	Waste management
Rabbani et al. [45]	2018	MILP	Heuristic	**Economic:** Minimize transport cost, **Environmental:** Minimize municipal waste impact	**Uncertain**	Waste management

J. Zhao and Ke [59]	2017	Bi-objective MILP	Heuristic	**Economic:** Minimize cost and risk	**Uncertain**	Waste management
Guo and Zhang [14]	2017	Integer programming	Heuristic	**Economic:** Minimize total cost	**Uncertain**	Reverse logistics for waste recycling
Farrokhi-Asl et al. [8]	2017	MILP	Metaheuristic	**Economic:** Minimize cost, **Social:** Maximize distance from disposal sites	**Certain**	Waste collection and distribution
Ghezavati and Beigi [11]	2016	Bi-objective programming	Metaheuristic	**Economic:** Minimize cost and time	**Uncertain**	Reverse logistics networks
V. F. Yu and Lin [56]	2016	MILP	Heuristic	**Economic**: Minimize total cost	**Certain**	Logistics and transportation
Kim and Lee [21]	2015	Integer programming	Heuristic	**Economic:** Minimize total costs	**Uncertain**	Reverse logistics
Khalili-Damghani and Naderi [20]	2014	ILP	Exact	**Economic:** Minimize total costs	**Certain**	Military logistics
Sheriff et al. [49]	2014	MINLP	Heuristic	**Economic:** Minimize total cost	**Certain**	Plastic recycling industry in India
S. Li et al. [24]	2013	MILP	Exact	**Economic:** Minimize total cost	**Certain**	Reverse logistics for food waste
H. Liu et al. [26]	2012	Multi-objective optimization	Heuristic	**Economic:** Minimize vehicles, Maximize network efficiency	**Uncertain**	Reverse logistics planning
Fernández	2006	Clarke-Wright VRP	Heuristic	**Economic:** Minimize total cost	**Certain**	Reverse logistics for mobile phones

factors are necessary if the goal is to have green and sustainable RL networks. Social goals such as consideration maximizing customer satisfaction (which is also a business goal), minimizing population risk, and maximizing the value of establishing potential sites are included in some studies. This factor is the least representative in the routing-location models for RL.

Certainty/Uncertainty: Specifies whether the model deals with certain or uncertain parameters. If any uncertain parameters exist in the model, then it is categorized as uncertain. Uncertainty is often related to fluctuating demands, costs, or capacities. Certain parameters assume that all parameters, such as costs and demand, are known and fixed. Even certain models can have parameters perturbed for sensitivity analysis. Application area: The specific context or industry where the model is applied, such as reverse logistics networks, waste management, e-commerce logistics, and bike-sharing systems with returns. Studies focus on design, optimization, and management of reverse logistics networks for various industries including e-commerce, medical waste, and bike-sharing systems. Thus, there are both service, manufacturing, and social use cases involved in these models. Several studies specifically target waste management, including municipal solid waste, medical waste, and industrial waste. Given that these wastes can serve as inputs for a circular system, and RL flows back into the forward supply chain, managing end-of-life materials is well represented. Some unique areas are also represented in these location-routing models that include relief logistics, e-commerce logistics, and military logistics, each with specific challenges and objectives. Some studies focus on broader supply chain management issues and include many of the other application areas. We alluded to circular economy practices such as circular supply selection (based on routing and location and other factors) and sustainable closed-loop supply chains.

This summary highlights the diversity in modeling approaches, solution techniques, and objectives in the field of reverse logistics and supply chain management, with a significant focus on economic efficiency, handling uncertainty, and application across various industries.

1.3.2 Critical and Specific Issues Identified Within the Literature

We begin by reviewing optimization models and studies that have used mathematical programming optimization for vehicle routing, location, and transportation within the overall reverse logistics network and in different settings. We will categorize the discussion in this section along categories of single objective/multiple objectives and whether they are certain or uncertain. We view these characteristics of the models can define and bound the approaches that will be used. Based on the set of publications in this review, we will see if there are significant differences in how studies address issues with these modeling characteristics in mind. There could have been many other categorizations that could have been used, but these characteristics are relatively mutually exclusive and provide an easy-to-understand framework. Thus, there are four subsections in this section: single objective with certainty, single objective with uncertainty, multiple objectives with certainty, and multiple objectives with uncertainty. Within each category, we will evaluate and synthesize our publication set around the following characteristics:

- What goals and objectives are most evident. Given that RL could be considered from economic, environmental, and social dimensions, we will determine the mix in each category.
- What RL activities are considered? Whether it is a full set or a specific activity of RL, or multiple activities will be determined based on the literature for this study.
- From a methodological perspective, we consider the predominance of solution techniques. Typically, the categories and considerations would be exact solution, heuristic, or multistage or hybrid approaches. Some of these issues will deal with complexity considerations.
- Finally, we will provide a quick evaluation on the type of industry, materials, or products managed in each model. We will now overview each of these categories of models and characteristics, with some exemplary citations included.

1.3.2.1 Single Objective—Certainty of Reverse Logistics Models

The single objective in this category of models most observed unsurprisingly is economic based. In each case, we found, and this is true for each case. The economic factor mentioned almost exclusively is to minimize cost. Economic objectives could also be based on increasing (maximizing) profit or revenue. This situation is not surprising for a few reasons. First, RL activities, in most cases, are not revenue generating but considered a cost center as are most transportation costs. Thus, there are activities that may only focus on companies seeking RL services as cost centers. Secondly, the focus is not on companies that provide these services where revenue (sales) of their services is applied. There is room for these types of objectives and services focus. A third reason is that application may be in nonprofit (government and military) settings [20]. We did not find any studies that considered multiple RL activities within this context of single objective with certainty. If these models were to consider storage and transformational activities, they would require more complexity to include production constraints (e.g., sorting, disassembly, and operational activities for recycling like crushing and processing). Modeling these multitier aspects with single objectives would need to consider tractability and purpose. Location analysis may be helpful, but vehicle routing between operational RL processers is relatively nonexistent in these models, although some studies that are more comprehensive exist in other more complex approaches discussed in the next subsections.

The RL activities considered in these routing and location problems were primarily associated with collection of materials. Vehicle routing and location analysis made collection and the relationship between companies and consumers a primary concern. The relationships and collection between organizations, a business-to-business perspective, were not as well developed for collection in these models. But some models do make a general consideration of customers as generic entities, which can include organizations [49].

Given that single-objective solutions may be more tractable, it was not surprising that at least one study sought out and introduced an exact solution [49]. But, in this situation, the model was solved exactly because of limited information that was provided by an actual case company which meant limiting the size of the problem. It is well known that many routing and location problems could be considered NP-hard, and the preponderance of these types of models calls for heuristic solution algorithms. Many are shown to be within 10% of exact solutions [2, 3].

The industry-materials-products applications of the single-objective certainty models were quite varied. As mentioned previously, one had a military-government application [20]. In this case, the objective was clearly aligned with specific requirements of repairing military equipment. The lack of variety in materials allowed this problem to be more easily modeled. The considerations here included economic-based materials and products that had incentives for their return and can be resold or reused (e.g., Hosseini et al., [16]). In addition to repair and reuse reasons, another circular economy practice reason is to manage the reverse logistics of food waste (end-of-life) materials [49]. Even in this small set of studies, we see that many types of return or collect reasons exist for investigating routing and location analysis whether they may be economic or environmentally related, even though the objective functions consider mostly economic concerns.

1.3.2.2 Single-Objective Uncertainty

The single-objective uncertainty models—like the certainty models—were mostly driven by economic goals and objectives. They included a variety of costs: most total cost measures included various transportation, location, and investment costs. Some of the costs may have been different in objective form, but the focus was still on costs. This characterization makes sense if there are only routing issues that need to be addressed rather than investments in building locations. One interesting study [60] used the term Fitness, rather than total cost, for the aggregation of costs in its model. We found no single objectives, even in more uncertain situations, with social and environmental measures, which may not be easy to evaluate and give greater uncertainties. Even in programs such as bike sharing, end-of-life or faulty bicycles, which have various environmental implications and consequences [48], consideration of environmental constraints and parameters is lacking.

The RL activities considered in these routing and location problems were primarily associated with collection of materials. But many studies with location and routing considered a dual-purpose aspect of collection. More than one study considered delivery and pickup (collection) simultaneously [36]. One study was unique in not only delivering to recycling centers [14], but also picking up products after recycling (the outbound logistics of recycled material) which is a relatively unique characteristic setting this study apart. These joint activities can be valuable for reverse logistics to make sure that transportation equipment is fully utilized with less energy and carbon emissions. Once again, some of these environmental characteristics were not incorporated into the single objective with uncertainty models.

Given the greater uncertainty in this set of models, it is likely that many solutions will not truly achieve optimality due to variations in the parameters and variables. Most techniques using various heuristic approaches such ant-bee colony (ABC) approaches were used. One of the techniques we consider exact and optimal uses column and constraint generation approaches [17]. They take advantage of assumptions and structure of their models (including a stochastic version of their deterministic model) and show using various instances that their optimality gap goes to 0. Most other papers in our data set did consider how well their heuristics performed to benchmarks and unsurprisingly their technique proved better. In this set of studies, the industries, products, and materials that were managed in the RL system were not given specific emphasis. In our limited data set, bike sharing and faulty bikes were one of the most specific applications [17]. Although one was clearly about recycling and end-of-life materials [14], the situation was relatively generic with no specific material or industry defined. The other studies focused less on the application and characterizations with a particular product or material and just focused on generic transportation networks. Modes of transportation, geographical situations, or other defining characteristics of the practical situation were missing. Thus, there is opportunity in greater application in practical settings for this category for RL modeling for routing, transportation, and location.

1.3.2.3 Multi-Objective Certainty

The multi-objective and certainty category has the fewest representative studies in this area of RL modeling for routing, location, and transportation analysis. There are a couple of potential reasons. First, with the inclusion of multiple objectives, it is likely that noneconomic environmental and social considerations will be included. These factors are usually difficult to model with certainty given the many intangible factors that are considered [37]. The other reason is that these models will also have many more factors, in general to consider. Just the increase in the number of factors will likely mean that there is more of a probability that these factors will have uncertain characteristics. Only one representative study was found in our search for RL routing, location, or transportation. Farrokhi-Asl et al. [8], the only paper in this category, includes economic and social dimensions but excludes explicit environmental objectives. The unique social dimension included is to minimize the social negativity of locating a waste facility (that would capture RL material flows) near certain locations. This locational objective is representative of the not in my backyard (NIMBY) concerns associated with end-of-life processing or waste locations. The RL activities covered in these studies both include a collection of a similar situation as most other studies from the previous two single-objective categories. No models have been considered for the additional stages of RL modeling. This is somewhat surprising, but it seems that the certainty characteristics of these models would preclude other RL activities, and there is room for further investigation along this line of thought to expand the considerations for other RL activities. Farrokhi-Asl et al. [9] address a bi-objective location-routing problem (LRP) for waste collection management by developing two metaheuristic algorithms: non-dominated sorting genetic algorithm II (NSGA-II) and multi-objective particle swarm optimization (MOPSO). This use of multiple heuristics is a common theme in many papers, especially given the NP-hard characteristics of vehicle routing and location. The study focused on a general description of the reverse logistics network and did not consider any specific industry or product. The model execution was relatively theoretical with simulated information. Given the limitations of this category of modeling, it seems that practical application would be limited.

1.3.2.4 Multi-Objective Uncertainty

This grouping of models was most represented in the literature on RL modeling routing, location, and transportation. The reason is because of the complexities involved in RL, especially when sustainability is a major driving force. The concerns are multifaceted and with many stakeholders that can be influenced by these decisions (e.g., not only companies and customers in the supply chain, but communities, regulators, and other third parties). There are very many perspectives involved in this model set. As in previous models, economic factors are quite prevalent and often with a number of cost (economic) and business objectives, such as risk [59]. Every model considered economic (business) factors as the exclusive model objective. Example objectives included total and "system" cost [57], travel time [11], and general "efficiency" [25]. Most of the previous objectives of the previous categories, and most of them in this category, focused on minimization issues, but some models included maximizing objectives. We mentioned efficiency already, and other maximizing objectives include client expectations [27], value of establishing sites [58], and customer (stakeholder) satisfaction [5, 6].

Another interesting aspect of the objectives is that social objective was the second most common sustainability objective, while environmental objective which is arguably a reason for many RL activities within the circular economy was only represented by two studies. The environmental objectives were to minimize total emissions and reduce environmental impacts [11, 16]. This surprising result for this modeling problem means there is substantial room for environmental and social dimensions to be explicitly modeled. No study included all three of the objectives with substantial room for more holistic sustainability evaluations for RL routing, location, and transportation evaluation models.

Overall, we will not go into detail here, and the solution methodologies were almost all heuristic or metaheuristic based. One of the few that could be considered "exact" had a unique focus on supplier selection [12]. It was also a multistage approach that used fuzzy data initially and translated the multi-objective model to a single-objective model that could be solved optimally. We quickly review a number of applications in this category. We provide a little more detail because of the unique motivations and characteristics of some of these applications and also the RL activities. One of the major regulatory policies and RL required consideration of loop closure involved product stewardship. An example of this are the Waste Electronics and Electrical Equipment (WEEE) regulations [38]. Companies had to manage the end of life of their products. Thus, optimizing vehicle routing in WEEE reverse logistics was needed [39]. The goal was not only to collect wastes, but also to enhance efficiency, manage uncertainties, reduce carbon emissions, and lower economic and environmental costs [39]. We mention this model explicitly due to its specific focus on meeting a specific application and regulatory policy. The regulatory policy has generated significant planning models beyond routing and location.

Urban logistics design has also played an important planning environment for RL [26]. A popular technique that we see in many RL routing and transportation studies includes mixed-integer linear programming (MILP) models, alone or with a hybrid optimization algorithm such as genetic algorithms (GGA) and variable neighborhood searches (GVNS). This approach, especially with multiple approaches, benefits reverse logistics by enabling comprehensive urban planning, integrating smart technologies for better reliability and sustainability, and reducing costs while improving service levels through optimized network design. Smart technologies have been rarely integrated, and we will recommend this area as a good direction for future research [51].

Another unique study application was for medical supplies [15]. In this case, operational efficiency, reduced logistics costs, and real-time planning adapt to complex environments and improve delivery speed and flexibility, especially in dynamic or urgent situations like medical supply distribution and waste collection. Real-time efficiency and solution techniques can prove valuable for sensitive RL activities. Applications also cover governmental, municipal developments [30], as evidenced by other category (single objective).

Recent studies also expanded the RL activities addressing collection-disassembly-delivery problem [22]. Newer techniques such as learning mechanisms have also been included. As a result of the application of these techniques, complex reverse logistics issues have been effectively addressed, particularly in scenarios involving uncertain product quality and return volume. As a result of the application of these techniques, complex reverse logistics issues have been effectively addressed, particularly in scenarios involving uncertain product quality and return volume. It has enabled manufacturers to reduce costs, improve customer satisfaction, and reduce environmental impact. Additionally, reverse logistics has been optimized, and processes have been streamlined using these techniques.

1.4 Future Reverse Logistics Modeling and Issues with Routing, Location, and Transportation

Managing the complexities of reverse logistics can also be accomplished through simulation, game theoretic, discrete alternative multiple criteria, and probabilistic predictive methodologies (to name a few). For example, using simulations, businesses can predict outcomes, identify bottlenecks, test various strategies, and identify shortcomings without disrupting actual operations on a large scale [35]. Discrete-event simulation can be effective in optimizing reverse logistics networks, particularly in addressing uncertainty and variability.

System dynamics simulations (SD), the optimization-based simulation (OBS) approach, can enhance waste electrical and electronic equipment (WEEE) management [29]. By doing so, system parameters can be updated periodically, thus improving responsiveness to market changes and increasing profitability and environmental benefits [29]. Inventory control problems can be integrated with simulation and optimization models [46]. The emphasis on inventory management within reverse logistics is intended to minimize costs and emphasize the importance of environmental sustainability. The field of reverse logistics is evolving rapidly, with emerging trends and future directions focusing on technological advancements and sustainability. Industry 4.0 technologies, such as the Internet of Things (IoT), and Digital Twin are transforming reverse logistics by enabling resource optimization, simulation and scenario analysis, predictive maintenance, enhanced product traceability, and real-time monitoring and scheduling of the supply chain. Technology adoption, however, occurs gradually over time and is a dynamic process. For reverse logistics operations to maximize benefits, this gradual change can affect key parameters. The long-term effects of technological upgrades can also alter the optimal configuration of reverse logistics networks.

The integration of low-carbon technologies into reverse logistics operations can also have substantial monetary savings in the long run, especially as low-carbon technologies become more affordable as they advance. Literature is still lacking, particularly when it comes to integrating these technologies with traditional reverse logistics practices. For more comprehensive solutions to reverse logistics challenges, future research should explore these areas.

1.5 Conclusion

Throughout this chapter, we focused on the optimization techniques that can enhance the efficiency and effectiveness of reverse logistics with a focus on routing and location analyses. We provided an overview of literature thematically evaluating the objectives, coverage of RL, applications, and solution techniques used. These are the major elements considered, but the richness of the field could have been examined from other perspectives, including theories used, outcomes, comparative analyses of when certain tools should be used, etc. We only provided a general overview of a very specific and limited set of issues that we feel are critical for RL planning and the circular economy. As observed in our analysis and discussion, significantly more research is needed expanding these models to a more holistic set of RL activities and more explicit consideration of sustainability concerns overall. These methodologies allow companies to address the complex challenges of reverse logistics, improve customer satisfaction, and contribute to environmental sustainability. Developing new technologies and approaches will be crucial as reverse logistics continues to evolve and becomes an integral component of sustainable supply chain management as the field continues to evolve.

References

1. Agrawal, S., Singh, R. K., & Murtaza, Q. (2015). A literature review and perspectives in reverse logistics. Resources, Conservation and Recycling (Vol. 97, p. 76–92). Elsevier. https://doi.org/10.1016/j.resconrec.2015.02.009
2. Bagheri Hosseini, M., Dehghanian, F., & Salari, M. (2019a). Selective capacitated location-routing problem with incentive-dependent returns in designing used products collection network. European Journal of Operational Research, 272(2), 655–673. https://doi.org/10.1016/j.ejor.2018.06.038
3. Bagheri Hosseini, M., Dehghanian, F., & Salari, M. (2019b). Selective capacitated location-routing problem with incentive-dependent returns in designing used products collection network. European Journal of Operational Research, 272(2), 655–673. https://doi.org/10.1016/j.ejor.2018.06.038
4. Banihashemi, T. A., Fei, J., & Chen, P. S.-L. (2019). Exploring the relationship between reverse logistics and sustainability performance. Modern Supply Chain Research and Applications, 1(1), 2–27. https://doi.org/10.1108/mscra-03-2019-0009
5. Beiki, H., Seyedhosseini, S. M., Ghezavati, V. R., & Seyedaliakbar, S. M. (2020a). Multi-objective optimization of multi-vehicle relief logistics considering satisfaction levels under uncertainty. International Journal of Engineering, Transactions B: Applications, 33(5), 814–824. https://doi.org/10.5829/IJE.2020.33.05B.13
6. Beiki, H., Seyedhosseini, S. M., Ghezavati, V. R., & Seyedaliakbar, S. M. (2020b). Multi-objective optimization of multi-vehicle relief logistics considering satisfaction levels under uncertainty. International Journal of Engineering, Transactions B: Applications, 33(5), 814–824. https://doi.org/10.5829/IJE.2020.33.05B.13
7. Bekrar, A., Cadi, A. A. El, Todosijevic, R., & Sarkis, J. (2021). Digitalizing the closing-of-the-loop for supply chains: A transportation and blockchain perspective. Sustainability (Switzerland), 13(5), 1–25. https://doi.org/10.3390/su13052895
8. Farrokhi-Asl, H., Tavakkoli-Moghaddam, R., Asgarian, B., & Sangari, E. (2017a). Metaheuristics for a bi-objective location-routing-problem in waste collection management. Journal of Industrial and Production Engineering, 34(4), 239–252. https://doi.org/10.1080/21681015.2016.1253619
9. Farrokhi-Asl, H., Tavakkoli-Moghaddam, R., Asgarian, B., & Sangari, E. (2017b). Metaheuristics for a bi-objective location-routing-problem in waste collection management. Journal of Industrial and Production Engineering, 34(4), 239–252. https://doi.org/10.1080/21681015.2016.1253619
10. Fleischmann, M., Bloemhof-Ruwaard, J. M., Dekker, R., Van Der Laan, E., Van Nunen, J. A. E. E., & Van Wassenhove, L. N. (1997). EUROPEAN JOURNAL OF OPERATIONAL RESEARCH Quantitative models for reverse logistics: A review. European Journal of Operational Rcsearch (Vol. 103).

11. Ghezavati, V. R., & Beigi, M. (2016). Solving a bi-objective mathematical model for location-routing problem with time windows in multi-echelon reverse logistics using metaheuristic procedure. Journal of Industrial Engineering International, 12(4), 469–483. https://doi.org/10.1007/s40092-016-0154-x
12. Govindan, K., Mina, H., Esmaeili, A., & Gholami-Zanjani, S. M. (2020). An Integrated Hybrid Approach for Circular supplier selection and Closed loop Supply Chain Network Design under Uncertainty. Journal of Cleaner Production, 242. https://doi.org/10.1016/j.jclepro.2019.118317
13. Govindan, K., Nosrati-Abarghooee, S., Nasiri, M. M., & Jolai, F. (2022). Green reverse logistics network design for medical waste management: A circular economy transition through case approach. Journal of Environmental Management, 322. https://doi.org/10.1016/j.jenvman.2022.115888
14. Guo, K., & Zhang, Q. (2017). A Discrete Artificial Bee Colony Algorithm for the Reverse Logistics Location and Routing Problem. International Journal of Information Technology and Decision Making, 16(5), 1339–1357. https://doi.org/10.1142/S0219622014500126
15. Han, Y., Zhang, M., Nan, P., Cao, J., Huang, Q., Ye, X., & He, Z. (2023). Two-stage heuristic algorithm for vehicle-drone collaborative delivery and pickup based on medical supplies resource allocation. Journal of King Saud University - Computer and Information Sciences, 35(10). https://doi.org/10.1016/j.jksuci.2023.101811
16. Hosseini Dehshiri, S. J., & Amiri, M. (2024). Considering the circular economy for designing closed-loop supply chain under hybrid uncertainty: A robust scenario-based possibilistic-stochastic programming. Expert Systems with Applications, 238. https://doi.org/10.1016/j.eswa.2023.121745
17. Huang, S., Liu, K., & Zhang, Z. H. (2023). Column-and-constraint-generation-based approach to a robust reverse logistic network design for bike sharing. Transportation Research Part B: Methodological, 173, 90–118. https://doi.org/10.1016/j.trb.2023.04.010
18. IEEE Systems, M., & Annual IEEE Computer Conference. ([s.d.]). 2019 6th International Conference on Control, Decision and Information Technologies (CoDIT´19) April 23-26, 2019, Le Cnam, Paris, France: conference digest.
19. Kenné, J. P., Dejax, P., & Gharbi, A. (2012). Production planning of a hybrid manufacturing remanufacturing system under uncertainty within a closed-loop supply chain. International Journal of Production Economics, 135(1), 81–93. https://doi.org/10.1016/j.ijpe.2010.10.026
20. Khalili-Damghani, K., & Naderi, H. (2014a). A mathematical location-routing model of repair centres and ammunition depots in order to support soldiers in civil wars. Int. J. Management and Decision Making (Vol. 13, Número 4).
21. Kim, J. S., & Lee, D. H. (2015). An integrated approach for collection network design, capacity planning and vehicle routing in reverse logistics. Journal of the Operational Research Society, 66(1), 76–85. https://doi.org/10.1057/jors.2013.168
22. Lei, J., Che, A., & Van Woensel, T. (2024). Collection-disassembly-delivery problem of disassembly centers in a reverse logistics network. European Journal of Operational Research, 313(2), 478–493. https://doi.org/10.1016/j.ejor.2023.07.008
23. Li, J., Zhao, Z., & Cheng, T. (2021). Research on LRP Integration of E-Commerce Logistics under the Background of Integration of Collection and Distribution. Discrete Dynamics in Nature and Society, 2021. https://doi.org/10.1155/2021/6038947
24. Li, S., Wang, L., Wu, H., & Bai, S. (2013). The location-routing problem in the food waste reverse logistics system. LISS 2012 - Proceedings of 2nd International Conference on Logistics, Informatics and Service Science, 149–154. https://doi.org/10.1007/978-3-642-32054-5_22
25. Liu, H., Zhang, Q., & Wang, W. (2011). Research on location-routing problem of reverse logistics with grey recycling demands based on PSO. Grey Systems, 1(1), 97–104. https://doi.org/10.1108/20439371111106768
26. Liu, Q., Chen, Y., Hu, W., Dong, J., Sun, B., & Cheng, H. (2023). Underground Logistics Network Design for Large-Scale Municipal Solid Waste Collection: A Case Study of Nanjing, China. Sustainability (Switzerland), 15(23). https://doi.org/10.3390/su152316392

27. Liu, S., Annual IEEE Computer Conference, IEEE International Conference on Grey Systems and Intelligent Services 2011.09.15-18 Nanjing, & GSIS 2011.09.15-18 Nanjing. ([s.d.]). IEEE International Conference on Grey Systems and Intelligent Services (GSIS), 2011 Nanjing, China, 15-18 September 2011.
28. Liu, S., Zhang, Y., Liu, Y., Wang, L., & Wang, X. V. (2019). An 'Internet of Things' enabled dynamic optimization method for smart vehicles and logistics tasks. Journal of Cleaner Production, 215, 806–820. https://doi.org/10.1016/j.jclepro.2018.12.254
29. Llerena-Riascos, C., Jaén, S., Montoya-Torres, J. R., & Villegas, J. G. (2021). An optimization-based system dynamics simulation for sustainable policy design in WEEE management systems. Sustainability (Switzerland), 13(20). https://doi.org/10.3390/su132011377
30. Luo, K., Zhao, W., & Zhang, R. (2024). A multi-day waste collection and transportation problem with selective collection and split delivery. Applied Mathematical Modelling, 126, 753–771. https://doi.org/10.1016/j.apm.2023.11.009
31. Luthra, S., Kumar Mangla, S., Kumar, S., Garg, D., & Haleem, A. (2017). Identify and prioritise the critical factors in implementing the reverse logistics practices: a case of Indian auto component manufacturer. Int. J. Business and Systems Research (Vol. 11, Número 2).
32. Macarthur, E. ([s.d.]). Founding Partners of the TOWARDS THE CIRCULAR ECONOMY Economic and business rationale for an accelerated transition.
33. Mahmoum Gonbadi, A., Genovese, A., & Sgalambro, A. (2021). Closed-loop supply chain design for the transition towards a circular economy: A systematic literature review of methods, applications and current gaps. Journal of Cleaner Production, 323. https://doi.org/10.1016/j.jclepro.2021.129101
34. Meade, L., Sarkis, J., & Presley, A. (2007). The theory and practice of Reverse Logistics. IEEE Transactions on Engineering Management, Engineering Management Journal, Supply Chain Management International Journal (Vol. 3, Número 1).
35. Min, H., Seong Ko, C., Ko, H.-J., Min, H., Ko, C., Ko, H., & Seong Ko is Professor, C. (2008). Managing product returns in customer-centric e-tail environments. Transportation Journal, Journal of Transportation Management and Transportation Research (Vol. 9, Número 1).
36. Nadizadeh, A., & Kafash, B. (2019). Fuzzy capacitated location-routing problem with simultaneous pickup and delivery demands. Transportation Letters, 11(1), 1–19. https://doi.org/10.1080/19427867.2016.1270798
37. Nandi, S., Hervani, A. A., Helms, M. M., & Sarkis, J. (2023). Conceptualising Circular economy performance with non-traditional valuation methods: Lessons for a post-Pandemic recovery. International Journal of Logistics Research and Applications, 26(6), 662–682. https://doi.org/10.1080/13675567.2021.1974365
38. Nowakowski, P., & Mrówczyńska, B. (2018). Towards sustainable WEEE collection and transportation methods in circular economy - Comparative study for rural and urban settlements. Resources, Conservation and Recycling, 135, 93–107. https://doi.org/10.1016/j.resconrec.2017.12.016
39. Oliveira Neto, G. C. de, de Araujo, S. A., Gomes, R. A., Alliprandini, D. H., Flausino, F. R., & Amorim, M. (2023). Simulation of Electronic Waste Reverse Chains for the Sao Paulo Circular Economy: An Artificial Intelligence-Based Approach for Economic and Environmental Optimizations. Sensors, 23(22). https://doi.org/10.3390/s23229046
40. Park, J., Sarkis, J., & Wu, Z. (2010). Creating integrated business and environmental value within the context of China's circular economy and ecological modernization. Journal of Cleaner Production, 18(15), 1494–1501. https://doi.org/10.1016/j.jclepro.2010.06.001
41. Pingping, Z., & Yu, Y. (2020). Research on Robust Bi-level Optimization of Reverse Logistics LRP under E- commerce Environment. Proceedings - 2020 International Symposium on Computer Engineering and Intelligent Communications, ISCEIC 2020, 9–13. https://doi.org/10.1109/ISCEIC51027.2020.00010
42. Plaza-Úbeda, J. A., Abad-Segura, E., de Burgos-Jiménez, J., Boteva-Asenova, A., & Belmonte-Ureña, L. J. (2021). Trends and new challenges in the green supply chain: The reverse logistics. Sustainability (Switzerland), 13(1), 1–19. https://doi.org/10.3390/su13010331

43. Pochampally, K. K., & Gupta, S. M. ([s.d.]). A Business-Mapping Approach to Multi-Criteria Group Selection of Collection Centers and Recovery Facilities. http://www.coe.neu.edu/~smgupta/
44. Rabbani, M., Amirhossein Sadati, S., & Farrokhi-Asl, H. (2020). Incorporating location routing model and decision making techniques in industrial waste management: Application in the automotive industry. Computers and Industrial Engineering, 148. https://doi.org/10.1016/j.cie.2020.106692
45. Rabbani, M., Mokhtarzadeh, M., & Farrokhi-Asl, H. (2018). A New Mathematical Model for Designing a Municipal Solid Waste System Considering Environmentally Issues. International Journal of Supply and Operations Management, 5(3), 234–255. www.ijsom.com
46. Rachih, H., Mhada, F. Z., & Chiheb, R. (2022). Simulation optimization of an inventory control model for a reverse logistics system. Decision Science Letters, 11(1), 43–54. https://doi.org/10.5267/j.dsl.2021.9.001
47. Rajput, S., & Singh, S. P. (2019). Connecting circular economy and Industry 4.0. International Journal of Information Management, 49, 98–113. https://doi.org/10.1016/j.ijinfomgt.2019.03.002
48. Saltykova, K., Ma, X., Yao, L., & Kong, H. (2022). Environmental impact assessment of bike-sharing considering the modal shift from public transit. Transportation Research Part D: Transport and Environment, 105. https://doi.org/10.1016/j.trd.2022.103238
49. Sheriff, K. M. M., Nachiappan, S., & Min, H. (2014). Combined location and routing problems for designing the quality-dependent and multi-product reverse logistics network. Journal of the Operational Research Society, 65(6), 873–887. https://doi.org/10.1057/jors.2013.22
50. Sun, S., Ma, L., Liu, Y., & Shang, C. (2023). Volleyball premier league algorithm with ACO and ALNS for simultaneous pickup–delivery location routing problem. Applied Soft Computing, 149. https://doi.org/10.1016/j.asoc.2023.111004
51. Sun, X., Yu, H., Solvang, W. D., & Govindan, K. (2024). A two-level decision-support framework for reverse logistics network design considering technology transformation in Industry 4.0: a case study in Norway. International Journal of Advanced Manufacturing Technology. https://doi.org/10.1007/s00170-024-14121-6
52. Tosarkani, B. M., Amin, S. H., & Zolfagharinia, H. (2020). A scenario-based robust possibilistic model for a multi-objective electronic reverse logistics network. International Journal of Production Economics, 224. https://doi.org/10.1016/j.ijpe.2019.107557
53. Trochu, J., Chaabane, A., & Ouhimmou, M. (2018). Reverse logistics network redesign under uncertainty for wood waste in the CRD industry. Resources, Conservation and Recycling, 128, 32–47. https://doi.org/10.1016/j.resconrec.2017.09.011
54. Walther, G., & Spengler, T. (2005). Impact of WEEE-directive on reverse logistics in Germany. International Journal of Physical Distribution and Logistics Management, 35(5), 337–361. https://doi.org/10.1108/09600030510607337
55. Wang, X. V., & Wang, L. (2019). Digital twin-based WEEE recycling, recovery and remanufacturing in the background of Industry 4.0. International Journal of Production Research, 57(12), 3892–3902. https://doi.org/10.1080/00207543.2018.1497819
56. Yu, V. F., & Lin, S. Y. (2016). Solving the location-routing problem with simultaneous pickup and delivery by simulated annealing. International Journal of Production Research, 54(2), 526–549. https://doi.org/10.1080/00207543.2015.1085655
57. Yuchi, Q., He, Z., Yang, Z., & Wang, N. (2016). A Location-Inventory-Routing Problem in Forward and Reverse Logistics Network Design. Discrete Dynamics in Nature and Society, 2016. https://doi.org/10.1155/2016/3475369
58. Zandkarimkhani, S., Nasiri, M. M., & Heydari, J. (2020). Sustainable open-loop supply chain network design considering location routing problem: A hybrid approach based on FAHP, FTOPSIS, and mathematical programming. International Journal of Logistics Systems and Management, 36(1), 92–123. https://doi.org/10.1504/IJLSM.2020.107229
59. Zhao, J., & Ke, G. Y. (2017). Incorporating inventory risks in location-routing models for explosive waste management. International Journal of Production Economics, 193, 123–136. https://doi.org/10.1016/j.ijpe.2017.07.001

60. Zhao, Y., Leng, L., & Zhang, C. (2021). A novel framework of hyper-heuristic approach and its application in location-routing problem with simultaneous pickup and delivery. Operational Research, 21(2), 1299–1332. https://doi.org/10.1007/s12351-019-00480-6
61. Zhou, J., Yang, S., Feng, H., & An, Z. (2023). Multi-echelon sustainable reverse logistics network design with incentive mechanism for eco-packages. Journal of Cleaner Production, 430. https://doi.org/10.1016/j.jclepro.2023.139500

Chapter 2
Designing a Closed-Loop Supply Chain Network Considering Uncertain Dynamic Demand and Production Quality

Taha-Hossein Hejazi, Parisa Cheshmeh, and Mohsen Shidpour

Abstract The closed-loop supply chain network design (CLSCND) involves integrating decision-making for both forward and reverse logistics. The significance of this design has increased due to various factors such as maximizing economic performance, competitive pressures, environmental concerns, resource utilization, and sustainability. In this study, we focus on a novel CLSCND that spans multiple periods and includes regional products. This research presents a mathematical model with two objectives: minimizing costs and minimizing environmental issues while focusing on imperfect quality production. This research begins by developing a mixed integer linear programming model to design a closed-loop supply chain network under uncertainty. The proposed model is then formulated based on robust optimization approaches to protect the network against uncertainty. Demand forecasting, specifically for nondeterministic dynamic demand, is addressed using the vector autoregressive time series model. Finally, numerical examples and sensitivity analysis on key parameters are provided to examine the behavior of the network's strength and validate the proposed model that offers valuable insights into effectively managing this problem. The results indicate that the unfulfilled demand significantly influences the overall cost of the supply chain.

Keywords Closed-loop supply chain · Environmental impacts · Uncertainty · Dynamic demand · Production quality · Robust optimization · Time series

T.-H. Hejazi (✉) · P. Cheshmeh
Department of Industrial Engineering, College of Garmsar, Amirkabir University of Technology (Tehran Polytechnic), Tehran, Iran
e-mail: t.h.hejazi@aut.ac.ir; Parisa.cheshmeh@aut.ac.ir

M. Shidpour
Department of Industrial Engineering, Amirkabir University of Technology (Tehran Polytechnic), Tehran, Iran
e-mail: Mohsen.shidpour@aut.ac.ir

B. Jarboui et al. (eds.), *Circular Economy Supply Chains*,
Computational Intelligence Methods and Applications,
https://doi.org/10.1007/978-981-96-7905-8_2

2.1 Introduction

Closed-loop supply chain network design (CLSCND) is a complex decision-making problem that aims to optimize multiple conflicting objectives, such as social, environmental, and economic aspects. It involves both forward and reverse supply chain activities, where the forward supply chain consists of suppliers, manufacturers, distribution centers, and customer areas working together to transform raw materials into finished products to meet customer demands. On the other hand, the reverse supply chain involves collection, recycling, and disposal centers that commence by collecting defective used products from customers. Remanufacturing, which involves recycling and reworking used and defective products into useful ones, offers a profitable approach to reducing the consumption of natural resources and waste generation. Remanufacturing involves recycling used and defective products to create new, useful products, offering a profitable solution to minimize the consumption of natural resources and waste production [1].

Supply chain network design (SCND) plays a crucial role in supply chain management, encompassing three levels of decision-making. The initial level involves strategic decisions, which have a lasting impact on supply chain performance and are costly and time-consuming to modify. These decisions involve selecting and determining the number of centers and facility capacity. The subsequent levels consist of tactical and operational decisions that are influenced by the strategic decisions made at the first level. These decisions offer more flexibility, can be easily adjusted to accommodate changes in the external environment, and optimize supply chain performance. However, it is important to note that SCND is a long-term decision that cannot be easily altered in response to short-term fluctuations in parameters. Neglecting the impact of uncertainty can significantly impact network configuration decisions. Therefore, it is essential for the supply chain to remain stable despite uncertain parameters.

Uncertainty in data is a common occurrence in problems, often caused by random factors, measurement errors, or other unknown reasons. To address this uncertainty, three approaches are commonly used: stochastic programming, fuzzy programming, and robust optimization. However, stochastic programming has a significant limitation in that it requires a sufficient amount of historical data to accurately determine the probability distribution of uncertain parameters. In cases where historical data is lacking, robust optimization (RO) can be a viable alternative to estimate the probability distribution of uncertain parameters.

This research examines a novel model for a multi-product multi-period closed-loop SCND. The model considers uncertainties in demand parameters, return rate, proportion of defective items, recycling costs, and demand forecasting error. Uncertain dynamic demand is projected through time series methodology (vector auto-regressive approach). To address these uncertainties, a robust optimization approach is employed.

The main contribution of this study is as follows:

- Imperfect products are produced due to the lack of high-quality production standards.
- Besides cost reduction in the supply chain network, this study focuses on reducing environmental footprints.
- Demand is viewed as a variable that is uncertain and constantly changing in this study.
- Demand forecasting is conducted through the utilization of time series methodology.

This research is structured as follows: Sect. 2.2 offers an overview of the existing literature on closed-loop supply chain design in the face of uncertainty and production quality within the supply chain. Section 2.3 outlines the research methodology, problem definition, development of the mathematical model, time series, and robust optimization approach. Section 2.4 presents the epsilon constraint method as the solution approach. Moving on to Sect. 2.5, we delve into the discussion of the model's results and validation and sensitivity analysis. Section 2.6 states managerial insights based on the outcomes of this study, and finally, in Sect. 2.7, we present our conclusions and outline potential avenues for future research.

2.2 Literature Review

Given that this investigation was derived from the CLSCND, taking into account the uncertain nature of dynamic demand and production quality, we analyzed the research conducted in two distinct domains.

- Closed-loop supply chain design under uncertainty
- Production quality in the supply chain

2.2.1 Closed-Loop Supply Chain Design Under Uncertainty

In recent years, numerous researchers have shown a keen interest in SCND. Today, the uncertainty of parameters stands as a significant challenge in designing an efficient supply chain network. This uncertainty arises due to the absence of precise and reliable information coupled with the intricate and dynamic nature of supply chain components. A recent study conducted by Govindan et al. [2] highlighted that over the past few decades, four primary strategies have been utilized to address uncertainty within the supply chain. These strategies include dynamic programming, stochastic programming, fuzzy programming, robust optimization, or any combination of these techniques.

Considering the uncertainty in the dynamic parameters of the model (e.g., market demand) represents a more realistic problem context. Consequently, there has been a recent focus on addressing random and dynamic demand. Additionally, fuzzy scheduling has gained popularity as an approach utilized by numerous researchers in the supply chain management domain to navigate uncertainty [3–9]. Stochastic programming is considered effective when the probability distribution of an uncertain parameter is known, serving as a crucial method to manage uncertainty in production planning and inventory control [10–12]. Salema et al. [13] introduced a stochastic mixed integer programming approach for multi-product networks in order to address demand uncertainty. Amin and Zhang [14] utilized a stochastic programming method to study the impact of demand uncertainty on supply chain network configuration. The network comprises multiple factories, collection centers, and demand markets. Their model aims to optimize total costs while maximizing the utilization of environmentally friendly materials.

Robust optimization is a cutting-edge methodology for addressing data uncertainty. Despite the absence of distributional assumptions regarding non-deterministic parameters, the data can still be effectively utilized. Several recent investigations [15–18] have adopted the stochastic robust optimization or scenario-based robust approach based on the approach introduced by Mulvey et al. [19].

Keyvanshokooh et al. [20] conducted a study on the CLSCND problem, focusing on uncertainties in demand, return rate, and transportation costs. They developed a hybrid stochastic-stochastic program aimed at maximizing expected profit. The network they analyzed consisted of various facilities such as manufacturing/remanufacturing centers, distribution centers, retailers, collection centers, and destruction centers. To solve the problem, they utilized the Accelerated Stochastic Benders Decomposition Algorithm. Ramezani et al. [21] proposed a robust optimization model for a multi-layer, multi-product, closed-loop supply chain network, considering uncertainties in demand parameters through a set of possible scenarios.

Recently, many scholars have been utilizing the robust optimization method to effectively handle uncertainties [22–30]. Talaei et al. [31] introduced a robust optimization approach specifically designed for the closed-loop supply chain network, with a primary focus on managing CO_2 emissions across the entire network. Their robust model successfully demonstrates its capability to effectively regulate network uncertainties. Prakash et al. [32] developed a novel integrated RO model for CLSCND in the presence of risk and demand uncertainty. They conducted a case study to validate the effectiveness of their model. Yavari and Geraeli [33] considered demand uncertainty, return rate, and the quality of returned products in the mathematical model. They proposed a robust optimization approach for CLSCND to handle uncertainty of the model. They also considered perishable product in the modeling approach.

2.2.2 Production Quality in the Supply Chain

Closed-loop supply chain models often assume that manufactured items are flawless. However, in reality, factors like machine deterioration, imperfect components, and raw materials can lead to the creation of defective products that must be taken into account during production. A few studies have been conducted on incorporating imperfect quality production into supply chain modeling, with the majority focusing on deterministic models [34–36]. Several researchers have also examined supply chain modeling under uncertainty, incorporating imperfect quality production into their studies [37]. Ramezani et al. [6] utilized fuzzy theory to develop a multi-period multi-product CLSCND. The supply chain structure in their study consisted of suppliers, factories, distribution centers, customers, collection centers, and disposal centers. The objective was to minimize the total number of defects in raw materials obtained from suppliers. This was achieved by selecting suppliers with superior raw materials, thereby enhancing the quality of the final products. Tiwari et al. [38] proposed an environmentally friendly production system that incorporated trade credits for customers. They also created a model for remanufacturing defective items. Furthermore, Maddah et al. [39] examined a production system where the produced/purchased items had two types of quality. The first type comprised items of excellent quality, while the second type consisted of items of imperfect quality. These imperfect items were identified during the inspection process but were not necessarily defective and were sold at a lower price.

Table 2.1 shows a summary of related studies. According to Table 2.1, demand, return rate, costs, and product quality are the most frequently cited sources of uncertainty, respectively. This study simultaneously considers demand, return rate, costs, demand forecasting error, and the ratio of defective items as additional sources of uncertainty. The robust optimization approach emerges as the most significant method for addressing uncertainty, particularly in situations where historical data is insufficient for estimating probability distributions of uncertain parameters. Furthermore, this study aims to simultaneously minimize supply chain costs and environmental impacts; as indicated in Table 2.1, only one paper has been found to align with these objectives.

2.3 Research Methodology

2.3.1 Problem Description and Model Development

The main goals of CLSCND are to enhance economic performance, optimize resource utilization, ensure high-quality delivery of goods to customers, address environmental and sustainability issues, gain competitive advantages, and boost social benefits by employing thorough decision-making in both forward and reverse logistics. When making important decisions in the supply chain, it is essential to

Table 2.1 Summary of literature review

Paper	Objective function			Structure				Uncertainty					Uncertainty approach				Supply chain attributes			
	Cost	Environmental	Other	Single Product	Multi-product	Single Period	Multi-period	Demand	Return Rate	Quality	Cost	Other	Stochastic robust optimization	Scenario-based robust approach	Random	Fuzzy	Closed-Loop Network Design	Transportation	Inventory	Other
[4]	•				•		•	•	•							•	•			
[6]	•		•		•		•	•	•		•					•	•	•	•	
[8]	•		•		•		•	•								•	•			
[9]	•	•	•		•		•	•				•				•	•			
[15]	•		•		•		•	•			•			•						•
[26]	•			•	•			•	•	•				•			•	•	•	
[28]	•		•	•		•		•			•		•					•		
[14]	•				•	•		•	•						•		•	•		
[40]	•			•		•				•					•		•			
[17]	•				•		•	•			•			•						•
[18]	•			•			•	•						•			•			
[20]	•			•			•	•	•		•			•			•			
[21]	•				•	•		•	•					•			•			
[23]	•				•	•		•				•	•					•	•	
[24]	•		•	•		•		•	•	•			•							
[25]	•				•	•		•			•	•	•				•			
[27]	•				•		•	•	•				•						•	•
[30]	•			•		•		•			•	•	•							
[32]	•			•		•		•				•	•				•			
[41]	•			•		•		•				•	•				•			•
[33]	•	•			•	•		•	•	•			•				•	•	•	•
This Research	•	•			•		•	•	•	•	•		•				•			

thoroughly assess uncertain factors such as demand, product return rates, costs, and product quality. Not considering demand uncertainty can result in anticipated losses in the long term for the supply chain, whereas enhancing short-term forecasting sets the stage for an efficient supply chain.

In this study, we consider a novel multi-period, multi-product, closed-loop supply chain network design. In this network, the manufacturing center is responsible for creating new, recycled, and remanufactured products. Subsequently, the manufactured products are dispatched to the distribution centers. At these facilities, the products undergo a thorough inspection and are subsequently stored before being transported to the predetermined and fixed customer zones based on the specific needs of each customer. Conversely, in the reverse logistics network, the collection centers, which are also responsible for inspection and inventory maintenance, retrieve the returned products from the customer's zones. Throughout the inspection process, products are categorized into two groups: defective products and used products. If the used items are deemed recyclable, they are forwarded to the manufacturer; otherwise, they are directed to the disposal center. Subsequently, defective items undergo a secondary inspection, with repairable products being returned to the manufacturing center for refurbishment, while non-repairable products are sent to disposal centers (Fig. 2.1).

The assumptions of the problem are as follows:

- Multi-stage, multi-product, and multi-layer supply chain is considered.
- Uncertain factors such as demand parameters, return rate, proportion of defective items, demand forecast error, recycling, and remanufacturing costs are taken into account.
- To address this uncertainty, a robust optimization approach is employed.
- The values of the uncertain parameters are treated as intervals.
- Time series analysis is utilized for demand forecasting.
- Some of the items produced are identified as defective.
- The customer zones are predetermined and remain fixed.
- The manufacturer is responsible for recycling and remanufacturing the products.

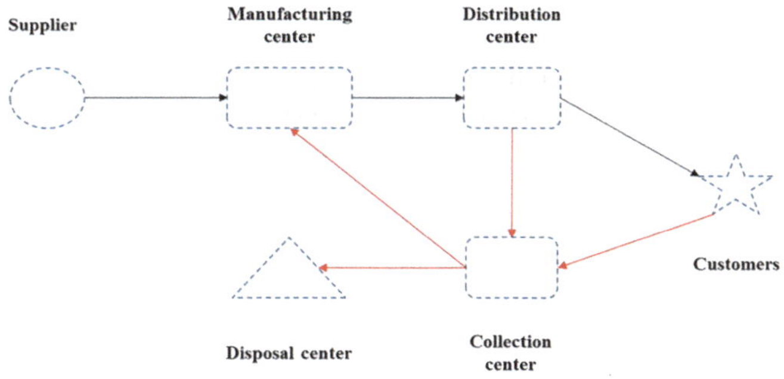

Fig. 2.1 The proposed CLSCND

- Besides facilitating product distribution, the distribution center also performs inspection and inventory holding.
- The collection center carries out product collection, inspection, and inventory maintenance.
- All centers in the supply chain have a capacity limit.

2.3.2 Proposed Methodology

This study aims to establish a framework for addressing the challenges associated with designing and implementing a closed-loop supply chain network under conditions of uncertain demand and production quality. The methodology employed in this research involves the application of operational research techniques, with a particular emphasis on mathematical modeling. The proposed approach focuses on optimization strategies to effectively manage the uncertainties inherent in the problem, such as demand fluctuation, return rates, defective item proportions, demand forecast inaccuracies, and costs associated with recycling and remanufacturing. Robust optimization is utilized to address the uncertainty of the problem, and the model is solved using the epsilon constraint method.

2.3.3 Mathematical Model

Sets

T : Set of periods indicated by the index t
S : Set of locations of supplier centers indicated by the index s
M : Set of locations of production centers indicated by the index m
I : Set of distribution center locations indicated by index i
C : Set of locations of the costumer zones indicated by the index c
L : Set of locations of collection centers indicated by index l
O : Set of locations of destruction centers indicated by index o
P : Set of products denoted by the index p
V : Set of vehicle types denoted by the index v

Nondeterministic parameters

$\widetilde{D}_{tpc}$:	Market demand for product p for customer region c in period t subject to uncertainty
$\widetilde{\theta}_{tp}$:	Return rate of product p in period t which is subject to uncertainty
$\widetilde{RC}_{tpm}$:	Recycling cost for product p at manufacturer m and period t subject to uncertainty
$\widetilde{REC}_{tpm}$:	Remanufacturing costs for product p at manufacturer m and in period t subject to uncertainty
$\widetilde{d}_t$:	The proportion of defective items in period t subject to uncertainty
$\widetilde{\varepsilon}_{tpc}$:	The error of forecasting demand for product p for customer area c and in period t subject to failure

Deterministic parameters

$\beta C1_p$:	Fraction of recyclable product p
$\beta C2_p$:	Fraction of the remanufactured product p sent from customer zone
$\beta d1_p$:	Fraction of remanufactured product p sent from a distribution center
FS_s :	Fixed cost of selecting suppliers s
FD_i :	Fixed cost of opening distribution center i
FC_l :	Fixed cost of opening collection center l
FO_o :	Fixed cost of opening disposal center o
SC_{ps} :	Per unit production cost for product p by supplier s
MC_{pm} :	Per unit production cost for product p by manufacturing center m
IC_{pi} :	Per unit inspection cost for product p at distribution center i
DC_{pi} :	Per unit processing cost of product p at distribution center i
Cc_{pl} :	Per unit collection cost for returned product p at collection center l
h_{pi} :	Inventory cost of product p at distribution center i
hw_{pl} :	Inventory cost of holding product p back to collection l
$\widehat{\pi}_{pc}$:	Shortage cost of product p for customer zone c
W_{pm} :	Order cost of product size p at manufacturing center m
P_{ps} :	Purchasing cost of product p from supplier s
Io_{po} :	Disposal cost of non-recyclable product p at the disposal center o
IL_{pl} :	Per unit inspection cost for product p at collection center l
CS_{ps} :	Raw material capacity of product p at supplier center s
CP_{pm} :	Production capacity of product p at manufacturing center m
CI_{pi} :	Product capacity p at distribution center i
CL_{pl} :	Product capacity p at collection center l
CO_{po} :	Product capacity p at disposal center o
CK_{pi} :	Storage capacity of product p in distribution center i
CW_{pl} :	Storage capacity of product p at collection center l
TMc_{psmv} :	Transportation cost of raw materials for product p from supplier s to manufacturing center m by vehicle v
TPc_{pmiv} :	Transportation cost of product p from manufacturing center m to distribution center i by vehicle v

TZc_{picv} : Transportation cost of product p from distribution center i to customer zone c by vehicle v

Tdl_{pilv} : Transportation cost of product p from distribution center i to collection center l by vehicle v

TOc_{pclv} : Transportation cost of product p from customer zone c to collection center l by vehicle v

$TOPc_{plmv}$: Transportation cost of product p from the collection center l to the manufacturing center m by vehicle v

TIc_{plov} : Transportation cost of product p from the collection center l to the disposal center o by vehicle v

$EIOL_l$: Environmental impact of opening collection center l

$EIOI_i$: Environmental impact of opening the distribution center i

$EIOO_o$: Environmental impact of opening the destruction center o

$EIHL_{pl}$: Environmental impact of keeping inventory of product p at collection center l

$EIHI_{pi}$: Environmental impact of keeping inventory of product p at the distribution center i

$EISS_{smpv}$: Environmental impact of transporting raw materials of product p from supplier s to manufacturing center m by vehicle v

$EISM_{mipv}$: Environmental impact of transporting product p from manufacturing center m to distribution center i by vehicle v

$EISI_{icpv}$: Environmental impact of transporting product p from distribution center i to customer zone c by vehicle v

$EISII_{ilpv}$: Environmental impact of transporting product p from distribution center i to collection center l by vehicle v

$EISC_{clpv}$: Environmental impact of transporting product p from customer zone c to collection center by vehicle v

$EISL_{lmpv}$: Environmental impact of transporting product p from collection center l to manufacturing center m by vehicle v

$EISII_{ilpv}$: Environmental impact of transporting product p from collection center l to disposal center o by vehicle v

Decision variables

QSM_{tpsmv} : The amount of raw material of product p ordered from supplier s to manufacturing center m in period t by vehicle v

QMD_{tpmiv} : The amount of product p shipped from manufacturing center m to distribution center i in period t by vehicle v

QDC_{tpicv} : Quantity of product p scheduled to be shipped from distribution center i to customer zone c in period t by vehicle v

QDL_{tpilv} : Amount of defective product p shipped from distribution center i to collection center l in period t by vehicle v

QNS_{tpc} : Amount of unsatisfied demand of product p for customer region c in period t

QCO_{tpclv} : The amount of product p returned from customer zone c to collection center l in period t by vehicle v

QRP_{tplmv} : Amount of recyclable product p sent from collection center l to manufacturing center m in period t by vehicle v

QEP_{tplmv} : Amount of remanufacturing product p sent from collection center l to manufacturing center m in period t by vehicle v

QIP_{tplov} : Amount of disposed product p sent from collection center l to disposal center o in period t by vehicle v

$nQDL_{tpi}$: Quantity of defective product p not shipped from distribution center i in period t

$nQIP_{tpl}$: Amount of waste product p not sent from collection center l in period t

$nQEP_{tpl}$: Amount of remanufacturing product p not shipped from collection center l in period t

$nQRP_{tpl}$: Amount of recyclable product p not shipped from collection center l in period t

SML_{plt} : Inventory level of product p at collection center l in period t

SMI_{pit} : Inventory level of product p at distribution center i in period t

v_{tpsm} : 1 if product p is ordered by manufacturing center m from supplier s in period t, otherwise 0

S_{ts} : 1 if the supplier is selected at location s in period t, 0 otherwise

DT_{ti} : 1 if the distribution center opened at location i in period t, otherwise 0

CT_{tl} : 1 if the collection center is opened at location l in period t, otherwise 0

DO_{to} : 1 if disposal center is opened at location o in period t, otherwise 0

$$
\begin{aligned}
Min\ Z_1 = & \sum_t \sum_s FS_s \cdot S_{ts} + \sum_t \sum_i FD_i \cdot DT_{ti} + \sum_t \sum_l FC_l \cdot CL_{tl} \\
& + \sum_t \sum_o FO_o \cdot DO_{to} + \sum_t \sum_p \sum_s \sum_m \sum_v W_{pm} \cdot v_{tpsm} \\
& + P_{ps} \cdot QSM_{tpsmv} \\
& + \sum_t \sum_p \sum_s \sum_m \sum_v QSM_{tpsmv} \left(SC_{ps} + TMC_{psmv} \right) \\
& + \sum_t \sum_p \sum_m \sum_i \sum_v QMD_{tpmiv} \left(MC_{pm} + TPC_{pmiv} + IC_{pi} \right) \\
& + \sum_t \sum_p \sum_i \sum_c \sum_v QDC_{tpicv} \left(DC_{pi} + TZC_{picv} \right) \\
& + \sum_t \sum_p \sum_c \sum_l \sum_v QCO_{tpclv} \left(Cc_{pl} + TOc_{pclv} \right) \\
& + \sum_t \sum_p \sum_l \sum_m \sum_v QRP_{tplmv} \left(\widetilde{RC}_{tpm} + TOPc_{plmv} \right) \\
& + \sum_t \sum_p \sum_l \sum_m \sum_v QEP_{tplmv} \left(\widetilde{REC}_{tpm} + TOPc_{plmv} \right) \\
& + \sum_t \sum_p \sum_l \sum_o \sum_v QIP_{tplov} \left(IO_{po} + TIc_{plov} \right) \\
& + \sum_t \sum_p \sum_c QNS_{tpc} \cdot \widehat{\pi}_{pc}
\end{aligned}
\tag{2.1}
$$

$$+\sum_t\sum_p\sum_i\sum_l\sum_v QDL_{tpilv}\cdot TDL_{pilv}$$
$$+\sum_p\sum_l\sum_t hw_{pl}\cdot SML_{plt}+\sum_p\sum_i\sum_t h_{pi}\cdot SMI_{pit}$$
$$+\sum_t\sum_p\sum_c\sum_l\sum_v\sum_i IL_{pl}\left(QCO_{tpclv}+QDL_{tpilv}\right)$$

The objective function (2.1) minimizes the total cost of the supply chain network. These costs include fixed costs for establishing supplier centers, distribution centers, collection centers, and disposal centers and ordering costs (if the manufacturer places an order for the supplier). Additionally, there are costs related to placing orders with suppliers, variable purchase costs, production and transportation costs incurred by supplier centers, manufacturing centers, distribution centers, collection centers, and disposal centers, costs related to recycling and remanufacturing, transportation costs, inspection costs, inventory cost, and shortage cost for when unmet demand is achieved in the network.

$$\begin{aligned}
Min\ Z_2 =& \sum_p\sum_l\sum_t EIHL_{pl}\cdot SML_{plt}+\sum_p\sum_i\sum_t EIHI_{pi}\cdot SMI_{pit}\\
&+\sum_t\sum_l EIOL_l\cdot CT_{tl}+\sum_t\sum_i EIOI_i\cdot DT_{ti}\\
&+\sum_t\sum_o EIOO_o\cdot DO_{to}\\
&+\sum_t\sum_p\sum_s\sum_m\sum_v EISS_{smpv}\cdot QSM_{tpsmv}\\
&+\sum_t\sum_p\sum_m\sum_i\sum_v EISM_{mipv}\cdot QMD_{tpmiv}\\
&+\sum_t\sum_p\sum_i\sum_c\sum_v EISI_{icpv}\cdot QDC_{tpicv}\\
&+\sum_t\sum_p\sum_c\sum_l\sum_v EISC_{clpv}\cdot QCO_{tpclv}\\
&+\sum_t\sum_p\sum_i\sum_l\sum_v EISII_{ilpv}\cdot QDL_{tpilv} \qquad (2.2)\\
&+\sum_t\sum_p\sum_l\sum_m\sum_v EISL_{lmpv}\cdot QRP_{tplmv}\\
&+\sum_t\sum_p\sum_l\sum_m\sum_v EISL_{lmpv}\cdot QEP_{tplmv}\\
&+\sum_t\sum_p\sum_l\sum_o\sum_v EISLL_{lopv}\cdot QIP_{tplov}
\end{aligned}$$

The objective function (2.2) minimizes environmental impacts. These effects are, respectively, inventory storage in collection centers and distribution centers, opening of collection, distribution and disposal centers, and transportation between centers (supplier to manufacturing center, manufacturing center to distribution center, distribution center to customer zones, customer to collection center, distribution center to collection center, collection center to manufacturing center, and collection to destruction centers).

$$\sum_{i}\sum_{v} QDC_{tpicv} + QNS_{tpc} \geq \tilde{D}_{tpc} \qquad \forall t \in T, p \in P, c \in C \tag{2.3}$$

Constraint (2.3) guarantees the customer's satisfaction of the unsatisfied demand.

$$\sum_{v}\sum_{l} QCO_{tpclv} \leq \widetilde{\theta}_{tp} \cdot \sum_{v}\sum_{i} QDC_{tpicv}\,, \qquad \forall t \in T, p \in P, c \in C \tag{2.4}$$

Constraint (2.4) states that not necessarily all the products adopted by the customer are returned to the collection centers.

$$\sum_{v}\sum_{m} QMD_{tpmiv} \cdot \left(1 - \tilde{d}_t\right) \geq \sum_{v}\sum_{c} QDC_{tpicv}\,, \quad \forall t \in T, i \in I, p \in P \tag{2.5}$$

Constraint (2.5) ensures that the quantity of healthy goods produced is greater than the quantity shipped to the customer's zone.

$$\sum_{v}\sum_{p}\sum_{c} (\beta C_P) \cdot QCO_{tpclv} \leq \sum_{v}\sum_{p}\sum_{o} QIP_{tplov}\,, \quad \forall t \in T, l \in L \tag{2.6}$$

Constraint (2.6) limits the amount of waste products transported from collection centers.

$$\sum_{v}\sum_{o} QIP_{tplov} + \sum_{v}\sum_{m} QRP_{tplmv} + \sum_{v}\sum_{m} QEP_{tplmv} = \sum_{v}\sum_{c} QCO_{tpclv}\,, \qquad \forall\, t \in T,\ p \in P, l \in L \tag{2.7}$$

$$\sum_{v}\sum_{o} QIP_{tplov} + \sum_{v}\sum_{m} QRP_{tplmv} + \sum_{v}\sum_{m} QEP_{tplmv} = \sum_{v}\sum_{c} QCO_{tpclv}\,, \qquad \forall\, t \in T,\ p \in P, l \in L \tag{2.8}$$

Constraints (2.7) and (2.8) show the flow balance between the layers.

$$\sum_{v} QSM_{tpsmv} \leq B \cdot v_{tpsm}\,, \qquad \forall\, t \in T, p \in P, s \in S, m \in M \tag{2.9}$$

Constraint (2.9) declares that the transfer of raw materials from the supplier to the manufacturer is only allowed once the contract between the two parties has been finalized.

$$QSM_{tpsmv},\ QMD_{tpmiv},\ QDC_{tpicv},\ QCO_{tpclv},\ QRP_{tplmv},\ QEP_{tplmv},$$
$$QIP_{tplov}, QNS_{tpc}, SML_{plt}, SMI_{pit}, QDL_{tpilv}, nQIP_{tpl}, nQRP_{tpl}, nQEP_{tpl},$$
$$nQDL_{tpi} \geq 0, \qquad \forall t \in T, p \in P, s \in S, m \in M, i \in I, l \in L, o \in O \tag{2.25}$$

$$DT_{ti},\ CT_{tl},\ DO_{to},\ v_{tpsm},\ S_{ts},\ \in \{0, 1\},$$
$$\forall t \in T, p \in P, s \in S, m \in M, i \in I, l \in L, o \in O \tag{2.26}$$

2.3.4 Forecasted Demand

In this research, the time series analysis approach is employed for demand forecasting, which is a quantitative method. This method aims to anticipate future data points by examining the observed data points within the series. In this particular study, the vector autoregressive method [42] is utilized to predict the demand starting from the second period. Consequently, the demand is defined by the following regression equation:

$$\tilde{D}_{tpc} = a_0 + \sum\nolimits_{z=1} \alpha_{zpc} \cdot D_{(t-z)pc} - \sum\nolimits_{j=1} b_{(t-j)pc} \cdot QNS_{(t-j)pc} + \widetilde{\varepsilon}_{tpc} \tag{2.27}$$

The equation mentioned above predicts the value $\tilde{D}_{tpc}$ based on the demand from the previous period. α_{zpc} and $b_{(t-j)pc}$ are parameters of the vector autoregressive method. Z and j denote the lag periods of the variables. $\widetilde{\varepsilon}_{tpc}$ represents the error between the actual and predicted results, with an uncertainty centered at zero. Based on the explanations provided above, constraint (2.3) applicable to each period is only taken into account for the initial period, resulting in

$$\sum\nolimits_i \sum\nolimits_v QDC_{1picv} + QNS_{1pc} \geq \tilde{D}_{1pc}, \qquad \forall p \in P, c \in C \tag{2.28}$$

Also, for the demand of later periods ($t \geq 2$), the following restriction will be added to the model:

$$\sum\nolimits_i \sum\nolimits_v QDC_{tpicv} + QNS_{tpc} \geq a_0 + \sum\nolimits_{z=1} \alpha_{zpc} \cdot D_{(t-z)pc}$$
$$- \sum\nolimits_{j=1} b_{(t-j)pc} \cdot QNS_{(t-j)pc} + \ \widetilde{\varepsilon}_{tpc}, \qquad \forall t \in T, p \in P, c \in C \tag{2.29}$$

2.3.5 Robust Optimization

The study utilized the robust optimization approach developed by Bertsimas and Sim [43, 44] to manage model uncertainty. This method allows for the adjustment of the level of conservatism. In the following, we apply this approach to the mathematical model.

$$
\begin{aligned}
Min\ Z_2 = &\sum_t \sum_s FS_s \cdot S_{ts} + \sum_t \sum_i FD_i \cdot DT_{ti} + \sum_t \sum_l FC_l \cdot CL_{tl} \\
&+ \sum_t \sum_o FO_o \cdot DO_{to} + \sum_t \sum_p \sum_s \sum_m \sum_v W_{pm} \cdot v_{tpsm} \\
&+ P_{ps} \cdot QSM_{tpsmv} \\
&+ \sum_t \sum_p \sum_s \sum_m \sum_v QSM_{tpsmv} \left(SC_{ps} + TMC_{psmv}\right) \\
&+ \sum_t \sum_p \sum_m \sum_i \sum_v QMD_{tpmiv} \left(MC_{pm} + TPC_{pmiv} + IC_{pi}\right) \\
&+ \sum_t \sum_p \sum_i \sum_c \sum_v QDC_{tpicv} \left(DC_{pi} + TZC_{picv}\right) \\
&+ \sum_t \sum_p \sum_c \sum_l \sum_v QCO_{tpclv} \left(Cc_{pl} + TOc_{pclv}\right) \\
&+ \sum_t \sum_p \sum_l \sum_m \sum_v QRP_{tplmv} \left(TOPc_{plmv}\right) \qquad (2.30) \\
&+ \sum_t \sum_p \sum_l \sum_m \sum_v QEP_{tplmv} \left(TOPc_{plmv}\right) \\
&+ \sum_t \sum_p \sum_l \sum_o \sum_v QIP_{tplov} \left(IO_{po} + TIc_{plov}\right) \\
&+ \sum_t \sum_p \sum_c QNS_{tpc} \cdot \widehat{\pi}_{pc} \\
&+ \sum_t \sum_p \sum_i \sum_l \sum_v QDL_{tpilv} \cdot TDL_{pilv} \\
&+ \sum_p \sum_l \sum_t hw_{pl} \cdot SML_{plt} + \sum_p \sum_i \sum_t h_{pi} \cdot SMI_{pit} \\
&+ \sum_t \sum_p \sum_c \sum_l \sum_v \sum_i IL_{pl} \left(QCO_{tpclv} + QDL_{tpilv}\right) \\
&+ y_{RC} + y_{REC}
\end{aligned}
$$

The objective function (2.19) incorporates the recycling and remanufacturing costs, which are subject to uncertainty. In order to effectively manage these costs, the objective function (2.30) is modified based on Almaraj and Trafalis' work [45]. This modified objective function includes the additional variables, y_{RC}, y_{REC}. These variables represent $QEP_{tplmv} \cdot \widetilde{REC}_{tpm}$ and $QRP_{tplmv} \cdot \widetilde{RC}_{tpm}$, respectively. Furthermore, the expressions are stated as constraints (2.31) to (2.34).

$$\sum_t \sum_p \sum_l \sum_m \sum_v QRP_{tplmv} \cdot \overline{RC}_{tpm} + \Gamma_{RC} U_{RC} + \sum_t \sum_p \sum_m P_{tpm} \leq y_{RC} \tag{2.31}$$

$$U_{RC} + P_{tpm} \geq \widehat{RC}_{tpm} \cdot \sum_l \sum_v QRP_{tplmv}, \qquad \forall t \in T, p \in P, m \in M \tag{2.32}$$

$$\sum_t \sum_p \sum_l \sum_m \sum_v QEP_{tplmv} \cdot \overline{REC}_{tpm} + \Gamma_{REC} U_{REC} + \sum_t \sum_p \sum_m P_{tpm} \leq y_{REC} \tag{2.33}$$

$$U_{REC} + P_{tpm} \geq \widehat{REC}_{tpm} \cdot \sum_l \sum_v QEP_{tplmv}, \qquad \forall t \in T, p \in P, m \in M \tag{2.34}$$

Due to the existence of the demand parameter, which is uncertain, Eq. (2.27) is written as:

$$\sum_i \sum_v QDC_{1picv} + QNS_{1pc} \geq \overline{D}_{1pc} + \Gamma_D \hat{D}_{1pc}, \qquad \forall p \in P, c \in C \tag{2.35}$$

Equation (2.29), due to an error parameter that has uncertainty, is transformed into:

$$\sum_i \sum_v QDC_{tpicv} + QNS_{tpc} \geq a_0 + \sum_z \alpha_{zpc} \cdot D_{(t-z)pc} - \sum_j b_{(t-j)pc} \cdot QNS_{(t-j)pc} + \overline{\varepsilon}_{tpc} + \Gamma_\varepsilon \widehat{\varepsilon}_{tpc}, \qquad \forall t \in T, p \in P, c \in C \tag{2.36}$$

The existence of the return rate parameter, characterized by uncertainty, leads to the establishment of Eq. (2.4) as Eqs. (2.37) and (2.38)

$$\sum_v \sum_l QCO_{tpclv} \leq \overline{\theta}_{tp} \cdot \sum_v \sum_i QDC_{tpicv} - \Gamma_\theta U_\theta - P_{tp}, \quad \forall t \in T, p \in P, c \in C \tag{2.37}$$

$$U_\theta + P_{tp} \geq \widehat{\theta}_{tp} \cdot \sum_v \sum_i \sum_c QDC_{tpicv} \qquad \forall t \in T, p \in P \tag{2.38}$$

Equation (2.5) is transformed into Eqs. (2.39) and (2.40) due to the presence of uncertainty in defective products parameter.

$$\sum_{v}\sum_{m} QMD_{tpmiv} \cdot \left(1 - \overline{d}_t\right) - \Gamma_d U_d - P_t \geq \sum_{v}\sum_{c} QDC_{tpicv}, \quad \forall t \in T, i \in I, p \in P \tag{2.39}$$

$$U_d + P_t \geq \hat{d}_t \cdot \sum_{p}\sum_{m}\sum_{i}\sum_{v} QMD_{tpmiv} \qquad \forall t \in T \tag{2.40}$$

Equation (2.15) is transformed into Eq. (2.41). In this constraint, the variable $nQDL_{tpi}$ is regarded as an adjustable variable.

$$nQDL_{tpi} - nQDL_{(t-1)pi} = \sum_{v}\sum_{m} QMD_{tpmiv} \cdot \overline{d}_t - \sum_{v}\sum_{l} QDL_{tpilv}, \qquad \forall t \in T, i \in I, p \in P \tag{2.41}$$

In this study, we developed a multi-product, multi-period CLSCND under uncertainty. Given the unpredictable nature of demand, we employed the time series method to forecast it. To address the uncertainties, such as demand, return rate, ratio of defective items, demand forecast error, recycling, and remanufacturing costs, we adopted a robust optimization approach. This research not only forecasts future demand data using time series analysis but also takes into account production with incomplete quality. Furthermore, in addition to optimizing the overall costs of the supply chain network, the research also focuses on reducing environmental impacts.

2.4 Solution Approach

This research develops a bi-objective model focusing on minimizing both costs and environmental impacts. To address this model, the epsilon constraint approach has been implemented, which is elaborated upon in the following.

The general form of a multi-objective decision-making (MODM) problem is as follows:

$$\begin{cases} Min\ \left(f_1(x), f_2(x), \ldots, f_n(x)\right) \\ x \in X \end{cases} \tag{2.42}$$

The epsilon constraint method focuses on achieving the main objective, such as minimizing costs, while limiting the secondary objectives, such as minimizing environmental impacts, within a specified upper limit called epsilon. These secondary objectives are transformed into constraints in the problem, resulting in a single-objective model.

$$\begin{cases} Min \ f_1(x) \\ f_i(x) \le e_i \quad i = 2, 3, \ldots, n \\ x \in X \end{cases} \tag{2.43}$$

To determine the suitable range for e_i related to the i-th goal ($i = 2, \ldots, n$), the initial step involves solving the subsequent optimization problems for each of the goals $j = 1, 2, \ldots, n$.

$$\begin{aligned} PayOff_{jj} &= Min \ f_j(x) \\ x &\in X \end{aligned} \tag{2.44}$$

The optimal solution is denoted as $x^{j,*}$, and the optimal value of the j-th objective is represented by $PayOff_{jj} = f_j\left(x^{j,*}\right)$. Consequently, the optimal value of objective i is obtained when any of the objectives $j = 1, 2, \ldots, n \ ; j \neq i$ is in the optimal state.

$$\begin{aligned} PayOff_{ij} &= Min \ f_i(x) \\ f_j(x) &= PayOff_{jj} \\ x &\in X \\ j &\neq i \end{aligned} \tag{2.45}$$

The optimal solution $x^{i,j,*}$ is determined by the optimal value of $PayOff_{jj} = f_i\left(x^{i,j,*}\right)$ for the i-th objective. Consequently, the payoff matrix is derived (Eq. (2.44)).

$$PayOff = \left[payOff_{ij}\right] \tag{2.46}$$

After the payoff matrix has been determined, the subsequent elements are established for the objective $i = 1, \ldots, n$.

$$Min(f_i) = Min_j\left\{payOff_{ij}\right\} = payOff_{ii} \tag{2.47}$$

$$Max(f_i) = Max_j\left\{payOff_{ij}\right\} \tag{2.48}$$

$$R(f_i) = Max(f_i) - Min(f_i) \tag{2.49}$$

With the above explanation, the appropriate interval for e_i is determined as $e_i \in [Min(f_i), Max(f_i)]$.

2.5 Computational Results

In this section, we assess the proposed model. Initially, to ensure its validity, the model is solved in six numerical examples using GAMS software with CPLEX solver and epsilon constraint method. Subsequently, the sensitivity analysis will be conducted on the key parameters of the model. All calculations are executed on a computer equipped with an Intel® Core i9 CPU, 64GB RAM, and 64-bit specifications. For the proposed problem, six scenarios were produced to analyze the problem in different sizes, and the results can be seen in Table 2.2. The solution time for the optimization model is based on the epsilon constraint method. The values of the parameters used in the proposed model are presented in the Appendix.

Figure 2.2 is based on a robust model with a high level of conservatism for scenario 1. Based on Fig. 2.2, there is a direct relation between lower costs and greater environmental impacts. For instance, if we aim to enhance the network's responsiveness, it would be necessary to reopen more centers to increase production or maintain higher inventory levels. These factors are the primary focus of this study due to their environmental implications. Managers can select the most suitable answer based on their specific goals and prevailing conditions.

Table 2.2 Scenarios

Scenarios	T	S	M	I	C	L	O	P	V	Execution time (s)
1	12	3	2	3	3	3	2	3	4	44.48
2	12	4	3	4	4	4	2	3	4	46.17
3	12	5	4	5	5	5	3	3	4	106
4	12	7	6	7	8	7	4	3	5	326
5	12	8	7	8	9	8	5	3	5	355
6	12	9	8	9	10	9	5	3	5	501

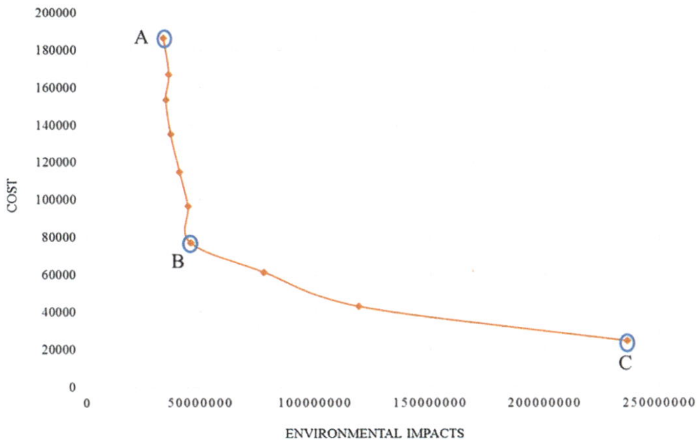

Fig. 2.2 Pareto front achieved by epsilon constraint method

As depicted in Fig. 2.2, point A exhibits the least environmental impact and the highest cost, while point C demonstrates the highest environmental impact and the lowest cost. The Pareto front indicates that points A and B show minimal disparity in environmental impacts, yet there is a substantial difference in costs between these two points. By referring to Fig. 2.2 and the Pareto front, the decision-maker can select the optimal solution based on their preference for each of the cost and environmental impact objectives.

2.5.1 Sensitivity Analysis

To assess the variations in objective function values and make informed strategic decisions, it is necessary to conduct a sensitivity analysis on key parameters that have an impact on the network. This analysis focuses on the changes in demand, proportion of defective items, shortage cost, inventory holding capacity, and budget. For this purpose, scenario 1 is utilized, considering the worst-case scenario (high conservatism) for robust optimization.

2.5.1.1 Sensitivity Analysis on Demand Parameter

In Table 2.3, it is evident that as demand increases, the quantity unsatisfied demand (QNS) also increases due to the constant number of centers in the supply chain network. Moreover, there are capacity constraints on centers, resulting in no change in production volume. For instance, in the base case, the first customer in the first period experiences unmet demand, as the demand is time-sensitive. Initially, there is unmet demand for two products, but with a 10% increase in demand, all three products face unmet demand. The shortage cost multiplies QNS in the objective function, significantly impacting the overall objective function value. Furthermore, according to constraint (2.35), the right side of the constraint grows as demand increases, resulting in a deterioration of the objective function. Conversely, a decrease in demand leads to a reduction in the right side of constraint (2.35), subsequently improving the value of the objective function.

Table 2.3 Sensitivity analysis on demand

Row	Demand changes	Objective changes	Number of suppliers	Number of distribution centers	Number of collection centers
1	10% increase	36% increase	Unchanged	Unchanged	Unchanged
2	20% increase	73% increase	Unchanged	Unchanged	Unchanged
3	10% decrease	35% decrease	Unchanged	Unchanged	Unchanged

Table 2.4 Sensitivity analysis on the ratio of defective products

Row	Defective items ratio changes	Objective changes	Number of suppliers	Number of distribution centers	Number of collection centers
1	50% increase	111% increase	Unchanged	Unchanged	Decrease
2	50% decrease	90.6% decrease	Decrease	Unchanged	Increase

2.5.1.2 Sensitivity Analysis on the Ratio of Defective Products

As depicted in Table 2.4, the customer's response capability diminishes as the ratio of defective items rises. Consequently, QNS increases, leading to a higher cost for the entire supply chain due to the elevated shortage cost. In accordance with constraint (2.39), an increase in the proportion of defective items results in a decrease in the proportion of healthy items. Since the left side of the constraint is greater than the right side, reducing it also reduces the left side of the constraint. Consequently, the distribution center dispatches a reduced quantity of products to the customer, thereby exacerbating the unmet demand. Furthermore, as the number of defective items rises, the distribution center dispatches a greater quantity of defective products to the collection centers. Decreasing the production volume leads to reduced orders from the manufacturer to the supplier. Consequently, the manufacturer utilizes the items received from the collection center to fulfill customer demand, and in order to improve responsiveness, the quantity of products returned from the customer to the collection center rises.

By decreasing the defective items ratio by 50%, the number of healthy items rises, prompting the distribution center to fulfill the customer's order quantity, resulting in zero QNS and the eradication of shortage costs. This leads to a significant 90.6% reduction in the overall chain costs. The decrease in defective items means fewer faulty products are sent to the collection center, causing an increase in the number of suppliers from 2 to 3 due to the high supplier selection costs and an increase in the volume of products shipped from the customers.

2.5.1.3 Sensitivity Analysis on Shortage Cost

The shortage cost has a significant impact on the value of the objective function, as it is multiplied by the unmet demand. However, it should be noted that increasing the shortage cost does not affect the variables indicated in Table 2.5.

2.5.1.4 Sensitivity Analysis on Inventory Capacity

As the capacity expands, both the inventory level of the distribution center and collection center experience a rise (Table 2.6). According to constraints (2.20) and (2.21), when the inventory level of the centers increases, the production volume

Table 2.5 Sensitivity analysis on shortage cost

Row	Shortage cost changes	Objective changes	Number of suppliers	Number of distribution centers	Number of collection centers
1	25% increase	22% increase	Unchanged	Unchanged	Unchanged
2	50% increase	46% increase	Unchanged	Unchanged	Unchanged
3	85% increase	77% increase	Unchanged	Unchanged	Unchanged

Table 2.6 Sensitivity analysis on inventory capacity

Row	Inventory capacity changes	Objective changes	Number of suppliers	Number of distribution centers	Number of collection centers
1	100% increase	91.3% decrease	Decrease	Unchanged	Decrease
2	50% decrease	167% increase	Decrease	Unchanged	Decrease

also increases. Consequently, the distribution center fulfills the customer's product demands and additionally provides more defective products to the collection center. This process continues until QNS reaches zero, resulting in a decrease in the overall cost of the entire supply chain. Due to high supplier selection costs, there has been a rise in the number of collection centers. Consequently, customers are dispatching a greater quantity of products, resulting in an increased flow of products from the collection center to both the manufacturer and the disposal center. By reducing capacity, the inventory levels at the distribution center and collection center are diminished. The production volume decreases, and it leads to a decrease in the number of products being sent from the distribution center to the customers, as well as a decrease in the number of defective products being sent to the collection center. As a result, the QNS has increased, raising the overall cost of the supply chain.

With the decrease in production, the manufacturer opts not to place an order with the supplier, leading to the earlier opening of the collection center in two periods. Consequently, as there is no supplier chosen to meet the demand, the collection center acquires more products from the customers and dispatches a greater quantity of products to both the manufacturer and disposal center.

2.5.1.5 Sensitivity Analysis of the Γ Parameter

Γ, also known as uncertainty budget, is another factor that impacts the objective function's value. To assess sensitivity analysis, this parameter's value is varied from 0 to 1 (as shown in Table 2.7).

Γ is an adjustable parameter that can be used to control the size of the uncertainty set. When Γ is set to zero, the problem is not safeguarded against uncertainty. Conversely, when Γ is 1, its maximum value, the problem is fully protected against uncertainty. For other values of Γ, the robustness of the problem can be adjusted based on the desired level of conservatism. As the Γ parameter increases, the

Table 2.7 Sensitivity analysis to the Γ parameter

Row	Γ value	Objective value	Number of suppliers	Number of distribution centers	Number of collection centers
1	0	1741343	Decrease	Decrease	Decrease
2	0.25	24275190	Unchanged	Unchanged	Unchanged
3	0.5	26640962	Unchanged	Unchanged	Unchanged
4	0.75	29724336	Unchanged	Unchanged	Unchanged
5	1	33596430	Unchanged	Unchanged	Unchanged

significance of uncertainty increases, leading to a higher level of conservatism, and more resources need to be allocated to the network. Consequently, by slightly augmenting the overall cost, the network can be protected against unpredictability.

As the Γ value increases, the unfulfilled demand also increases. In line with constraint (2.35), the higher the Γ value, the higher the customer demand. Similarly, in constraint (2.39), the left side of the constraint decreases, leading to a reduction in the quantity of healthy products produced by the manufacturing center. Consequently, the distribution center also delivers a smaller amount of products to the customers, thereby reducing the responsiveness of the chain. Based on the constraint (2.37), an increase in Γ leads to a reduction in the right side of the constraint. Consequently, the return rate of products decreases, resulting in a decrease in the quantity of products collected by the collection center and a subsequent decline in the level of inventory. Referring to Table 2.7, a decrease in the Γ parameter corresponds to a decrease in the number of centers due to the reduction in uncertainty.

2.6 Managerial Insights

Based on the findings from the sensitivity analysis discussed earlier, with the aim of minimizing supply chain costs, an increase in unfulfilled demand is an unfavorable outcome for the supply chain network. The rise in product demand directly corresponds to an escalation in unmet demand. Therefore, it is crucial to address this issue by implementing specific strategies.

Managers may consider resolving this issue by enhancing the inventory holding capacity of the facilities, given the cost-effectiveness of inventory management. This strategy helps prevent shortages and mitigate the rise in chain costs. With an expanded capacity, manufacturers can raise production volumes, and collection centers can gather more products. This boost in capacity has the potential to drive increased production and sales, ultimately fostering job creation within the supply chain. This proposed approach contradicts the goal of minimizing environmental impacts due to the fact that inventory maintenance contributes to the rise in environmental impacts. As a result, managers have the option to forward the products that were not dispatched to the distribution and collection center in the preceding period to the subsequent period in order to prevent any escalation in their environmental footprint. Furthermore, decision-makers have the ability to enhance the level of customer response during the initial phase and foster trust by improving the quality of their products. This, in turn, results in a decrease in the proportion of defective products, subsequently reducing transportation expenses and minimizing waste percentage. As the second period commences, decision-makers can meet customers' demands by utilizing returned products, which incur significantly lower costs for recycling and remanufacturing compared to production costs. These changes can be significant due to the decreased utilization of natural resources and the reduction of the detrimental impacts stemming from CO_2 emissions. According to the results of this study that there is a direct relationship between lower costs and higher environmental effects, managers should consider trade-offs between costs and environmental effects in order to make optimal decisions according to business conditions.

Companies can reduce the conservatism level of the supply chain network by forecasting parameters with a degree of uncertainty, thus preventing extra expenses. Furthermore, there is no necessity to open additional centers in order to decrease the pollution generated by the operation of these facilities.

The increase in unmet demand can also be attributed to the rise in defective items within the supply chain, resulting in higher costs. To address this issue, managers can designate a portion of the produced items as potentially defective and optimize their operations accordingly. By doing so, they can avoid the need for major changes in the network or strategic decisions, ensuring that customer needs are met without any shortages. Manufacturers can reduce the number of defective products used in their production process by implementing strategies such as sourcing materials from trustworthy suppliers, conducting regular maintenance and repairs on machinery, and implementing inspection stations at the production site. By doing so, they enhance the overall quality of their output and prevent incurring additional expenses associated with defective items, such as inspection fees, transportation costs, and inventory maintenance costs, as well as support environmental issues.

2.7 Conclusion and Future Research

Managers are increasingly focused on minimizing costs and environmental impacts. As a result, companies are grappling with the task of creating a closed-loop supply chain that effectively addresses both of these objectives.

In this research, a mixed integer linear programming model is developed to design a closed-loop supply chain network under demand uncertainty, return rate, demand forecast error, defective item ratio, and recycling and remanufacturing costs. The proposed model incorporates two objective functions, minimizing both costs and environmental impacts. The epsilon constraint approach is employed to solve this model. The model is developed within the context of imperfect production and is structured utilizing the robust optimization methodology to safeguard the network from uncertainties. In this problem, non-deterministic dynamic demand is considered, which is forecasted through the utilization of a time series model.

Based on the results conducted for the given issue, six scenarios of different sizes were generated, and a sensitivity analysis was carried out on the critical parameters of the problem. The results of the sensitivity analysis indicated that the unfulfilled demand significantly influences the overall cost of the supply chain. This influence may stem from an upsurge in demand, a rise in the percentage of defective products, an increase in the gamma parameter, and a reduction in inventory capacity.

Our proposed model is adaptable to a range of industries, encompassing steel production companies, electronics industries, and automobile manufacturing centers, where returned products (whether defective or previously utilized) can be reused as primary resources.

For future studies, the following can be mentioned:

- Maximizing social benefits and achieving sustainability
- Considering more uncertainties such as the quality of returned products
- Forecasting significant parameters such as the proportion of defective items and the quality of returned products using time series and regression methods
- Development of heuristics and meta-heuristics solution methods for larger dimensions of the problem

References

1. Liu, C., Zhu, Q., Wei, F., Rao, W., Liu, J., Hu, J., Cai, W.: An integrated optimization control method for remanufacturing assembly system. Journal of Cleaner Production **248**, 119261 (2020)
2. Govindan, K., Fattahi, M., Keyvanshokooh, E.: Supply chain network design under uncertainty: A comprehensive review and future research directions. European journal of operational research **263**(1), 108–141 (2017)
3. Shekarian, E., Kazemi, N., Abdul-Rashid, S.H., Olugu, E.U.: Fuzzy inventory models: A comprehensive review. Applied Soft Computing **55**, 588–621 (2017)
4. Pishvaee, M.S., Torabi, S.A.: A possibilistic programming approach for closed-loop supply chain network design under uncertainty. Fuzzy sets and systems **161**(20), 2668–2683 (2010)
5. Demirel, N., Özceylan, E., Paksoy, T., Gökçen, H.: A genetic algorithm approach for optimising a closed-loop supply chain network with crisp and fuzzy objectives. International journal of production research **52**(12), 3637–3664 (2014)
6. Ramezani, M., Kimiagari, A.M., Karimi, B., Hejazi, T.H.: Closed-loop supply chain network design under a fuzzy environment. Knowledge-Based Systems **59**, 108–120 (2014)
7. Fallah, H., Eskandari, H., Pishvaee, M.S.: Competitive closed-loop supply chain network design under uncertainty. Journal of Manufacturing Systems **37**, 649–661 (2015)
8. Amin, S.H., Baki, F.: A facility location model for global closed-loop supply chain network design. Applied Mathematical Modelling **41**, 316–330 (2017)
9. Soleimani, H., Govindan, K., Saghafi, H., Jafari, H.: Fuzzy multi-objective sustainable and green closed-loop supply chain network design. Computers & industrial engineering **109**, 191–203 (2017)

10. Masih-Tehrani, B., Xu, S.H., Kumara, S., Li, H.: A single-period analysis of a two-echelon inventory system with dependent supply uncertainty. Transportation Research Part B: Methodological **45**(8), 1128–1151 (2011)
11. Wang, D., Qin, Z., Kar, S.: A novel single-period inventory problem with uncertain random demand and its application. Applied Mathematics and Computation **269**, 133–145 (2015)
12. Zhang, Z., Li, Y., Huang, G.: An inventory-theory-based interval stochastic programming method and its application to Beijing's electric-power system planning. International Journal of Electrical Power & Energy Systems **62**, 429–440 (2014)
13. Salema, M.I.G., Barbosa-Povoa, A.P., Novais, A.Q.: An optimization model for the design of a capacitated multi-product reverse logistics network with uncertainty. European journal of operational research **179**(3), 1063–1077 (2007)
14. Amin, S.H., Zhang, G.: A multi-objective facility location model for closed-loop supply chain network under uncertain demand and return. Applied Mathematical Modelling **37**(6), 4165–4176 (2013)
15. Mirzapour Al-E-Hashem, S., Malekly, H., Aryanezhad, M.B.: A multi-objective robust optimization model for multi-product multi-site aggregate production planning in a supply chain under uncertainty. International journal of production economics **134**(1), 28–42 (2011)
16. Ruimin, M., Lifei, Y., Maozhu, J., Peiyu, R., Zhihan, L.: Robust environmental closed-loop supply chain design under uncertainty. Chaos, Solitons & Fractals **89**, 195–202 (2016)
17. Rahmani, D., Ramezanian, R., Fattahi, P., Heydari, M.: A robust optimization model for multi-product two-stage capacitated production planning under uncertainty. Applied Mathematical Modelling **37**(20-21), 8957–8971 (2013)
18. Safaei, A.S., Roozbeh, A., Paydar, M.M.: A robust optimization model for the design of a cardboard closed-loop supply chain. Journal of cleaner production **166**, 1154–1168 (2017)
19. Mulvey, J.M., Vanderbei, R.J., Zenios, S.A.: Robust optimization of large-scale systems. Operations research **43**(2), 264–281 (1995)
20. Keyvanshokooh, E., Ryan, S.M., Kabir, E.: Hybrid robust and stochastic optimization for closed-loop supply chain network design using accelerated benders decomposition. European journal of operational research **249**(1), 76–92 (2016)
21. Ramezani, M., Bashiri, M., Tavakkoli-Moghaddam, R.: A robust design for a closed-loop supply chain network under an uncertain environment. The International Journal of Advanced Manufacturing Technology **66**, 825–843 (2013)
22. Aharon, B.T., Boaz, G., Shimrit, S.: Robust multi-echelon multi-period inventory control. European Journal of Operational Research **199**(3), 922–935 (2009)
23. Baghalian, A., Rezapour, S., Farahani, R.Z.: Robust supply chain network design with service level against disruptions and demand uncertainties: A real-life case. European journal of operational research **227**(1), 199–215 (2013)
24. Hatefi, S.M., Jolai, F.: Robust and reliable forward–reverse logistics network design under demand uncertainty and facility disruptions. Applied mathematical modelling **38**(9–10), 2630–2647 (2014)
25. Kisomi, M.S., Solimanpur, M., Doniavi, A.: An integrated supply chain configuration model and procurement management under uncertainty: a set-based robust optimization methodology. Applied Mathematical Modelling **40**(17–18), 7928–7947 (2016)
26. Pishvaee, M.S., Rabbani, M., Torabi, S.A.: A robust optimization approach to closed-loop supply chain network design under uncertainty. Applied mathematical modelling **35**(2), 637–649 (2011)
27. Wei, C., Li, Y., Cai, X.: Robust optimal policies of production and inventory with uncertain returns and demand. International Journal of Production Economics **134**(2), 357–367 (2011)
28. Cao, J.X., Li, X.X., Wang, Z.Y., Wu, J.: The robust model of continuous transportation network design problem with demand and cost uncertainties. Procedia-Social and Behavioral Sciences **96**, 970–980 (2013)
29. Zhang, Y., Jiang, Y.: Robust optimization on sustainable biodiesel supply chain produced from waste cooking oil under price uncertainty. Waste Management **60**, 329–339 (2017)

30. Zokaee, S., Jabbarzadeh, A., Fahimnia, B., Sadjadi, S.J.: Robust supply chain network design: an optimization model with real world application. Annals of Operations Research **257**, 15–44 (2017)
31. Talaei, M., Moghaddam, B.F., Pishvaee, M.S., Bozorgi-Amiri, A., Gholamnejad, S.: A robust fuzzy optimization model for carbon-efficient closed-loop supply chain network design problem: a numerical illustration in electronics industry. Journal of cleaner production **113**, 662–673 (2016)
32. Prakash, S., Kumar, S., Soni, G., Jain, V., Rathore, A.P.S.: Closed-loop supply chain network design and modelling under risks and demand uncertainty: an integrated robust optimization approach. Annals of operations research **290**, 837–864 (2020)
33. Yavari, M., Geraeli, M.: Heuristic method for robust optimization model for green closed-loop supply chain network design of perishable goods. Journal of Cleaner Production **226**, 282–305 (2019)
34. Rad, M.A., Khoshalhan, F., Glock, C.H.: Optimal production and distribution policies for a two-stage supply chain with imperfect items and price-and advertisement-sensitive demand: A note. Applied Mathematical Modelling **57**, 625–632 (2018)
35. Masoudipour, E., Amirian, H., Sahraeian, R.: A novel closed-loop supply chain based on the quality of returned products. Journal of cleaner production **151**, 344–355 (2017)
36. Sana, S.S.: A production-inventory model of imperfect quality products in a three-layer supply chain. Decision support systems **50**(2), 539–547 (2011)
37. Hu, J.S., Zheng, H., Xu, R.Q., Ji, Y.P., Guo, C.Y.: Supply chain coordination for fuzzy random newsboy problem with imperfect quality. International Journal of Approximate Reasoning **51**(7), 771–784 (2010)
38. Tiwari, S., Ahmed, W., Sarkar, B.: Multi-item sustainable green production system under trade-credit and partial backordering. Journal of cleaner production **204**, 82–95 (2018)
39. Maddah, B., Salameh, M.K., Karame, G.M.: Lot sizing with random yield and different qualities. Applied Mathematical Modelling **33**(4), 1997–2009 (2009)
40. Jeihoonian, M., Zanjani, M.K., Gendreau, M.: Closed-loop supply chain network design under uncertain quality status: Case of durable products. International Journal of Production Economics **183**, 470–486 (2017)
41. Kim, J., Do Chung, B., Kang, Y., Jeong, B.: Robust optimization model for closed-loop supply chain planning under reverse logistics flow and demand uncertainty. Journal of cleaner production **196**, 1314–1328 (2018)
42. Wang, C.H., Chen, J.Y.: Demand forecasting and financial estimation considering the interactive dynamics of semiconductor supply-chain companies. Computers & Industrial Engineering **138**, 106104 (2019)
43. Bertsimas, D., Sim, M.: Robust discrete optimization and network flows. Mathematical programming **98**(1), 49–71 (2003)
44. Bertsimas, D., Sim, M.: The price of robustness. Operations research **52**(1), 35–53 (2004)
45. Almaraj, I.I., Trafalis, T.B.: An integrated multi-echelon robust closed-loop supply chain under imperfect quality production. International Journal of Production Economics **218**, 212–227 (2019)

Chapter 3
Applying a Simulated Annealing Embedded Genetic Algorithm for Solving a Concurrent Stochastic Supply Chain and Product Family Design with the Location-Allocation of Retailers

Taha-Hossein Hejazi, Fatemeh Jahangir, Ebrahim Rezaee Nik, and Sepideh Asadi Zeidabadi

Abstract Product family design with supply chain design within the framework of a circular economy adds more complexity to meeting diverse demands. This chapter investigates a product family design problem in a stochastic three-echelon supply chain, including a manufacturer, retailers, and customers under demand uncertainty. Also, two groups of customers are considered that the first seeks cheaper products, and the second seeks higher quality products. So, it is essential to determine the quality level and cost of products so that both groups of customers are satisfied. In this research, both mandatory and arbitrary modules are incorporated into product design, with customers selecting the arbitrary modules. This approach enhances product variety based on customer preferences, minimizes waste from unnecessary materials and components, and supports sustainable production practices and a circular economy. Considering the variety of demands in product family problems, the retailers should be located so that the transportation cost is reduced and the demands are satisfied, simultaneously. In order to increase the efficiency of resources and reduce environmental effects, a limit has been considered for serving customers, in which each customer is only allowed to buy from retailers who are present in the customer's area. The objective of the proposed model is defined

T.-H. Hejazi (✉)
Department of Industrial Engineering, College of Garmsar, Amirkabir University of Technology (Tehran Polytechnic), Tehran, Iran
e-mail: t.h.hejazi@aut.ac.ir

F. Jahangir · E. R. Nik
Department of Industrial Engineering, Sadjad University of Technology, Mashhad, Iran
e-mail: rezaeenik@sadjad.ac.ir

S. A. Zeidabadi
Department of Industrial Engineering and Management Systems, Tehran, Iran
e-mail: sepideh.asadi@aut.ac.ir

B. Jarboui et al. (eds.), *Circular Economy Supply Chains*,
Computational Intelligence Methods and Applications,
https://doi.org/10.1007/978-981-96-7905-8_3

product family design problem. They also introduced a bi-level joint optimization model for leader-follower in a multi-objective programming space to concurrently decide about product family architecture and assembly configuration. Their research led to a reduction in assembly complexity and costs of design and assembly. Luo et al. [15] presented choosing an intelligent production line and a product family design problem so that the greenhouse gas emission was reduced. They concluded using the intelligent production line enhances the profit of the firm.

3.2.2 *Quality in the Supply Chain*

Product family design seeks to meet customer demand for customized products as well as reduce supply chain costs. But another important issue is providing high-quality products to customers, which adds complexity to the product family problem.

Franca et al. [16] designed a bi-objective four-echelon supply chain model to simultaneously increase quality and profit. They defined quality as minimizing the amount of defective items in supplied raw materials. Also, they considered costs, demand, inventory level, and volume of products as stochastic. Yin and Nishi [17] formulated a three-echelon supply chain model considering the number of defective items and uncertain demand. They considered asymmetric quality data among the manufacturing plant and supplier. In this research, the manufacturer is responsible for supplier selection when the supplier quality information is unknown. Nagurney et al. [18] explored the rivalry between the manufacturer and the freight service regarding pricing and quality. Different transportation modes and asymmetric demand were also considered in this research. Hejazi et al. [36] worked on a supply chain model in which different production technologies, waste reduction strategies, inspection policies, and cost minimization were regarded. In this research, the producer can choose the production technology based on both efficiency and costs to reduce waste items. Hejazi and Khorshidvand [20] investigated product family design in a closed-loop supply chain, regarding costs, quality level, air pollution, and missed working days under demand uncertainty. They employed the robust optimization approach against uncertainty. Mridha et al. [21] proposed a three-echelon sustainable supply chain aimed at reducing carbon emissions, enhancing the quality of biofuel, and minimizing waste. They adopted a two-stage inspection policy to separate impure biofuels and dispose of them as waste. They applied an algebraic approach to maximize the profit of the production plant. Huma et al. [22] studied the connection between supply chain quality integration and green supply chain and then evaluated their effects on the triple bottom line. The supply chain quality integration propels the suppliers to use eco-friendly and high-quality products. They realized that supply chain quality integration increases both the green buying activities and the performance of internal environmental management.

3.2.3 Uncertainty

In most real optimization cases, data is usually fluctuating. As mentioned before, four methods exist to reduce the impact of data fluctuation and uncertainty. The product family design process is a complex problem due to the diversity of modular products, but this problem becomes more complex when some data in the supply chain, such as customer demand, is uncertain.

Baud-Lavigne et al. [23] applied robust optimization in the simultaneous design of supply chain and product family where customer demand is unknown. They incorporated upgrading the product quality and functionality, the possibility of buying modules from subcontractors, and changing the sequence of operations in their model. Hejazi and Seyyed-Esfahani [24] studied a multistage quality optimization problem where quality in each stage is affected by several different design factors. They employed the multi-response method due to multiple inputs and outputs. They developed a stochastic scenario-based programming model to minimize deviations of the outputs and quality losses. Homayounmehr and Hejazi [25] introduced a four-echelon supply chain model considering costs, reliability, demand, and location capacity, in which the bi-stage stochastic programming was used due to the demand uncertainty. In this research, the distribution of stochastic parameters was defined by Bernoulli. Gómez-Rocha and Hernández-Gress [26] employed stochastic programming to deal with the uncertainties in production capacity and demand in a multistage aggregate production planning model considering two product families. They assumed the probability distribution of random variables to be normal. Foroozesh et al. [27] employed generalized interval type-2 fuzzy to optimize disruption and operational risks of durable and perishable items in a resilient and green supply chain under epistemic uncertainties. They were able to increase the usage of facility capacities and decrease network costs and air pollution. Zarouk et al. [28] introduced a simultaneous scheduling and vehicle routing problem within a green supply chain regarding stochastic demands. Also, different transportation times between nodes due to drivers' tiredness, soft service horizons, and ceiling of permitted driving time for each customer in a heterogeneous transportation system were investigated. They applied the GASA to solve their model and compared its results with the exact method. Zeidabadi et al. [4] investigated cost minimization and workforce efficiency maximization in a supply chain where demand, costs, and backorder were assumed to be uncertain. They applied scenario-based robust optimization to deal with uncertainties. Wang et al. [29] introduced a stochastic transportation problem for perishable products where the demands are determined after the products arrive at the customers' location. They considered two types of vehicles to respond to demands: The first type is allocated to the initial needs of customers, and the second one is dedicated to unfulfilled demands. They mixed adaptive large neighborhood search and NSGA-II to seek Pareto optimal responses where the first objective was minimizing operating cost and the second was maximizing customers' satisfaction. Bayram et al. [30] worked on a stochastic hub network optimization and dynamic routing problem considering uncertain demands, capacity acquisition, and congestion. They concluded that the use of congestion costs and uncertain demand would increase hub capacity levels.

3.2.4 Contribution

The major features of this study are mentioned below:

- A product family design problem with two types of customers is considered: The first group is more sensitive to cost, and the second group is more sensitive to the quality of products.
- A product design problem with integrating both mandatory and arbitrary modules is considered in which customers are allowed to choose the arbitrary components. This increases product variety to meet customer preferences, reduces waste of surplus materials, and promotes sustainable production and a circular economy.
- To enhance resource efficiency and minimize environmental impact, a restriction has been implemented whereby each customer may only purchase from retailers located within their own area.
- A three-echelon supply chain is considered, including a manufacturer, retailers, and markets.
- A simultaneous supply chain and product family design is developed in which the retailers are located so that transportation cost is minimized and the variety of demands is met.
- The cost elements, including fixed development of modules, transportation, locating of retailers, and production, are considered in the model.
- Both locating and allocating retailers to the markets have been considered.
- A stochastic supply chain considering demand uncertainty is developed.
- Considering the NP-hard structure of the proposed model, the GASA algorithm is employed to solve the model.

According to the literature review and Table 3.1, few studies have investigated the role of the location of retailers in the product family problems, while it affects the transportation cost and fulfilling various demands significantly. So in this research, we have studied the effect of the transportation policies, demand uncertainty, and the location of retailers and allocation them to the markets on a simultaneous supply chain and product family design. Also, in product family design problems, modules and options can be selected according to the customer's preference over cost or quality. As a result, a stochastic three-echelon supply chain and modular product design problem under demand uncertainty are developed.

Table 3.1 Comparison between previous research and current study

Paper	Product family	Quality	Finished product		Location	Transportation	Approach	Uncertain parameter
			Single	Multiple				
Wu et al. [7]	✓	✓	✓				Deterministic	
Franca et al. [16]		✓		✓		✓	Stochastic	Demand profit
Das and Chowdhury [8]	✓	✓		✓		✓	Stochastic	Demand
Yin and Nishi [17]		✓		✓			Stochastic	Demand
Nagurney et al. [18]		✓		✓		✓	Deterministic	
Rezapour et al. [2]	✓	✓		✓	✓		Deterministic	
Ma and Kremer [9]	✓		✓			✓	Fuzzy	Product end of life
Aguila et al. [10]	✓			✓			Deterministic	
Gan et al. [11]	✓		✓				Deterministic	
Ziaei et al. [12]	✓			✓			Deterministic	
Homayounmehr and Hejazi [25]					✓		Stochastic	Demand
Gómez-Rocha and Hernández-Gress [26]	✓			✓			Stochastic	Demand
Foroozesh et al. [27]	✓			✓	✓	✓	Fuzzy and robust	Capacity of facilities-demand-transportation cost-CO2 emission
Zarouk et al. [28]				✓		✓	Stochastic	Demand-supply
Hejazi and Khorshidvand [20]	✓	✓		✓		✓	Robust	Demand
Hossain et al. [13]	✓			✓			Deterministic	
Zeidabadi et al. [4]				✓		✓	Robust	Demand-backorder-costs
Wang et al. [29]		✓	✓			✓	Stochastic	Demand
Bayram et al. [30]				✓	✓	✓	Stochastic	Demand
Luo et al. [15]	✓			✓			Deterministic	
This study	✓	✓		✓	✓	✓	Stochastic	Demand

3.3 Proposed Model

3.3.1 Problem Description

Quality and price are two important criteria when a customer buys a product. The quality and price of the product are determined according to the quality and price of the modules. In this research, customers are distributed into two categories: the first group pays more attention to the price, and the rest to the quality. Each product consists of different modules that are chosen by the customer. These modules must be compatible with each other and be able to respond to the price and quality desired by the customer. Two types of modules, including mandatory and arbitrary, are considered in this research. Arbitrary modules improve the performance of products, so they have more cost than mandatory modules. The better the quality of the modules, the higher the production cost and the final price of the product [2]. This research investigates a family product problem in a three-echelon supply chain, considering profit, quality of products, locating and allocating retailers to the markets, and transportation planning. Furthermore, in this research, both mandatory and arbitrary modules are employed in product design to boost product variety to align with customer preferences, minimize waste from excess materials and parts, and support sustainable production and a circular economy. Besides, to improve resource efficiency and reduce environmental impact, customers are restricted to purchasing only from retailers within their local area.

3.3.2 Assumptions

The proposed model assumptions are defined as follows:

- ✓ The proposed model has been developed as a bi-product and single-period model.
- ✓ A three-echelon supply chain model regarding a manufacturer, multiple retailers, and multiple customers is developed.
- ✓ Suppliers supply all materials, and the manufacturer cannot produce the materials.
- ✓ Customer demand is considered stochastic and quantified by random scenarios.

3.3.3 Mathematical Modeling

In this section, first, the nomenclature is explained, and then the mathematical model is explained:

- ***Sets and Indices***

i	Index for mandatory or arbitrary modules
j	Index for existing modules options
k	Index for existing markets of products
r	Index for available retailers
n	Index for customers' categories
ss	Index for scenarios
m	Set of mandatory modules
f	Set of arbitrary modules
m_j^m	Set of existing options j for mandatory module m
f_j^f	Set of existing options j for arbitrary module f
$NM^{m_j^m}$	Set of incompatible options with mandatory option m_j^m existing in the other modules
$NF^{f_j^f}$	Set of incompatible options with arbitrary option f_j^f existing in the other modules
cu	Set of customers' categories

- ***Parameters***

A_{ss}^k	Demand of market k under scenario ss
E_n	Minimum product order capacity for the nth category of customers
mp_n	Expected profit per unit of product allocated to the nth category of customers
qq_{cu}	Minimum product quality level for category of customers cu
$qM_m^j(qF_f^j)$	Quality level for the jth option of mandatory module m (arbitrary module f)
$CM_m^j(CF_f^j)$	Production cost per unit of the jth option of mandatory module m (arbitrary module f)
$DM_m^j(DF_f^j)$	Fixed development cost per unit of the jth option of mandatory module m (arbitrary module f)
Sc_n^k	Total supply cost for the nth category of customers in market k
cc^r	Transportation cost for each unit of product from the factory to retailer$r$
L_r^k	Location cost of retailer r in market k
B_n^k	Percentage of the nth category of customers in market k
rr_n^k	Sensitivity of the nth category of customers to the quality of productions in the market k
G_n^k	Sensitivity of the nth category of customers to the price of productions in the market k
pi_{ss}	The occurrence probability of scenario ss
BM	A big constant number

- ***Variables***

$x_i^{j,n}$ 1 if the jth option of module i (mandatory or arbitrary) is chosen for the designed product for the nth category of customers; 0 otherwise

z_f^n 1 if the arbitrary module f is embedded in the designed product for the nth category of customers; 0 otherwise

hx_m^j 1 if the mandatory module m_j^m is selected for at least one product; 0 otherwise

hv_f^j 1 if the arbitrary module f_j^f is selected for at least one product; 0 otherwise

w_r^k 1 if the retailer r located in market k; 0 otherwise

q_n The designed product quality for the nth category of customers

mc_{ss}^n Total cost of producing the designed product for the nth category of customers under scenario ss

$p_{n,ss}^k$ Price of retailing the designed product for the nth category of customers in the market k under scenario ss

$o_{n,ss}^{k,r}$ Quantity of dispatched demand for the nth category of customers in market k by retailer r under the scenario ss

The mathematical formula of the proposed model is described below:

$$\begin{aligned}
\text{Max z} = & \left(\sum_{ss} pi_{ss} \sum_{n}\sum_{k}\sum_{r} mp_n o_{n,ss}^{k,r}\right) \\
& - \left(\sum_{m}\sum_{j} DM_m^j hx_m^j + \sum_{f}\sum_{j} DF_f^j hv_f^j\right) \\
& - \left(\sum_{ss} pi_{ss} \sum_{n}\sum_{k}\sum_{r} cc^r o_{n,ss}^{k,r}\right) - \left(\sum_{k}\sum_{r} L_r^k w_r^k\right). \qquad (3.1)
\end{aligned}$$

Equation (3.1) denotes maximizing the expected profit in the objective function. The first segment in the objective function shows the profit of the supply chain for various markets. In the second section, the fixed development costs of all modules are illustrated. The third section depicts the transportation cost per unit of product from the factory to the retailer. Finally, the fourth section indicates the location cost of each retailer. Equations (3.2–3.18) denote the constraints of the model.

$$o_{n,ss}^{k,r} \leq w_r^k BM \qquad \forall r, k, n, ss. \qquad (3.2)$$

Equation (3.2) ensures that each product can only be distributed in those markets where a retailer is located.

$$o_{n,ss}^{k,r} \leq (A_{ss}^{k} B_{n}^{k}) + rr_n^k \left(\sum_m \sum_j qM_m^j x_m^{j,n} + \sum_f \sum_j qF_f^j x_f^{j,n} \right)$$
$$- G_n^k \left(\sum_m \sum_j CM_m^j x_m^{j,n} + \sum_f \sum_j CF_f^j x_f^{j,n} + Sc_n^k + mp_n \right)$$
$$+ BM(1 - w_r^k) \qquad \forall r, k, n, ss. \tag{3.3}$$

Equation (3.3) demonstrates that the dispatched products to the retailers are less than or equal to the demand.

$$\sum_j x_m^{j,n} = 1 \qquad \forall m, n. \tag{3.4}$$

Equation (3.4) guarantees that one of the options of the mandatory module must be embedded in the product structure.

$$\sum_j x_f^{j,n} = z_f^n \qquad \forall f, n. \tag{3.5}$$

Equation (3.5) guarantees the existence of one of the options of the arbitrary module if it is embedded in the designed product structure.

$$x_t^{s,n} \leq BM(1 - x_m^{j,n}) \qquad \forall m_t^s \in NM^{m_j^m}. \tag{3.6}$$

Equation (3.6) reveals incompatibility between the mandatory options.

$$x_t^{s,n} \leq BM(1 - x_f^{j,n}) \qquad \forall m_t^s \in NF^{f_j^f}. \tag{3.7}$$

Equation (3.7) depicts incompatibility between the arbitrary options.

$$\sum_k \sum_r o_{n,ss}^{k,r} \geq E_n \qquad \forall n, ss. \tag{3.8}$$

Equation (3.8) illustrates that the demand must be higher than the minimum product order capacity.

$$x_m^{j,n} \leq hx_m^j \qquad \forall m, j, n. \tag{3.9}$$

Equation (3.9) shows that a mandatory option can be selected only when the corresponding mandatory module is selected.

$$hx_m^j \leq \sum_m \sum_j x_m^{j,n} \qquad \forall m, j, n. \tag{3.10}$$

Equation (3.10) ensures that if a mandatory option is embedded in the structure of the designed product, the corresponding mandatory module must also exist in the structure of the designed product.

$$x_f^{j,n} \leq hv_f^j \qquad \forall f, j, n. \tag{3.11}$$

Equation (3.11) shows that an arbitrary option can be selected only when the corresponding arbitrary module is selected.

$$hv_f^j \leq \sum_f \sum_j x_f^{j,n} \qquad \forall f, j, n. \tag{3.12}$$

Equation (3.12) ensures that if an arbitrary option is embedded in the structure of the designed product, the corresponding arbitrary module must also exist in the structure of the designed product.

$$q_n = \sum_m \sum_j qM_m^j x_m^{j,n} + \sum_f \sum_j qF_f^j x_f^{j,n} \qquad \forall n. \tag{3.13}$$

Equation (3.13) calculates the quality of the designed products considering both mandatory and arbitrary modules.

$$q_n \geq qq_{cu} \qquad \forall n, cu(n \neq cu). \tag{3.14}$$

Equation (3.14) demonstrates the minimum quality level of designed products.

$$mc_{ss}^n = \sum_m \sum_j CM_m^j x_m^{j,n} + \sum_f \sum_j CF_f^j x_f^{j,n} \qquad \forall n, ss. \tag{3.15}$$

Equation (3.15) calculates the total cost of production of the designed products regarding both mandatory and arbitrary modules.

$$p_{n,ss}^k = \sum_m \sum_j CM_m^j x_m^{j,n} + \sum_f \sum_j CF_f^j x_f^{j,n} + Sc_n^k + mp_n \qquad \forall k, n, ss. \tag{3.16}$$

Equation (3.16) indicates the retail price of the designed products which is obtained through the sum of costs and expected profit of the firm.

$$\sum_r w_r^k = 1 \qquad \forall k. \tag{3.17}$$

Equation (3.17) reveals that each market is supplied by a retailer.

$$\sum_{k} w_r^k \leq 1 \qquad \forall r. \tag{3.18}$$

Equation (3.18) guarantees that each retailer is allocated to a maximum of one market.

This section described a problem of stochastic mixed-integer linear programming for the product family. The superiority of this model compared to previous works is as follows:

✓ The customers can choose between the cheaper products, which naturally have lower quality levels, and the higher quality products, which are logically more expensive.
✓ The concurrent supply chain and product family design considering the location-allocation of the retailers was developed.
✓ The fixed development cost, transportation cost, location cost, and production cost were included in the model.
✓ A stochastic model under demand uncertainty was expanded.

3.4 Computational Results

In this section, first, a numerical instance is expanded to solve the model. Next, a sensitivity analysis is employed to analyze the impact of the model changes on results. Then a dimensional analysis is executed to examine the behavior of the solution time in different dimensions of the problem. For this purpose, the CPU times of the exact method and a metaheuristic approach are compared. The Taguchi method is also applied to specify the best levels of parameters in the metaheuristic approach. Finally, the comparison results between the exact method and the metaheuristic approach are demonstrated.

A three-echelon supply chain network is developed considering a manufacturer, retailers, and customers. Each product contains several modules, and each module contains two types of options. The first option has a higher quality and, therefore, a higher cost; the second option has a lower cost and, therefore, a lower quality. Also, two types of customers are considered in this chapter: the first group pays more attention to the product price, and the second one pays more attention to the product quality. A set of data considering 11 mandatory modules, three arbitrary modules, three options, six markets, 10 retailers' locations, two end products, and 15 scenarios has been generated to solve the model. The probability of each scenario is obtained using the discrete uniform distribution function, which is calculated as 1/(a total number of scenarios) $= 1/15$.

In the numerical instance, all parameters have been produced using uniform distribution except customer demand generated by triangular distribution [2]. The information on the distribution of parameters is given in the Appendix. Note that each unit of cost is equivalent to 1000 dollars. The proposed model described in Sect. 3.3.3 is solved by GAMS software [31] to obtain exact solutions on a computer with Intel(R) Core i5 2450M CPU and 2.5 GHz processor with 6.00 GB RAM, and a sensitivity analysis is developed to evaluate the proposed model. Tables 3.2 and 3.3 show the results of solving the model for product 1 and product 2, respectively. The highlighted cells in Tables 3.2 and 3.3 show the chosen options.

3.4.1 Sensitivity Analysis

In this section, we intend to employ sensitivity analysis to examine the effect of changing parameters on solutions. Considering that one group of customers pays more attention to cost and the second group gives more importance to quality, we examine the impact of changes in cost and quality on customer choices. The sensitivity analysis can help reduce useless materials and promote circularity by evaluating how changes in prices affect customer behavior. In essence, it identifies which options of arbitrary modules are preferred by customers when product prices fluctuate. The findings indicate that customers generally favor the second option among various choices. Therefore, it is advisable to offer smaller quantities or avoid providing additional options to enhance circularity and reduce excess materials.

For this purpose, first, we increased the quality level of the product 1 from 20 to 50. As can be seen in Table 3.4, with the increasing sensitivity of customers to the product's quality level, modules with higher quality levels are selected.

Considering the increase in the price of product 2 from 20 to 50, it can be seen in Table 3.5 choosing arbitrary modules or higher quality modules is reduced.

If we reduce the costs of the modules, the probability of choosing an arbitrary module for product 2 increases because by reducing costs, the final price of the product will also decrease. For this purpose, we changed the fixed development cost from $uniform(20, 30)$ to $uniform(10, 15)$ and presented the results of selected modules in Table 3.6.

Suppose we increase the costs of the modules. In that case, the probability of choosing an arbitrary module for product 2 decreases because, by increasing costs, the product's final price would also increase. For this purpose, we changed the production cost of each unit of product 2 from $uniform(0.03, 0.05)$ to $uniform(0.06, 0.1)$ and presented the results of selected modules in Table 3.7.

Figure 3.1 shows the solution of the proposed model for the first scenario considering retailers' optimal locations. In Figs. 3.1 and 3.2, the numbers in each box above the bars from left to right indicate the market number, the allocated retailer to that market, and the amount of fulfilled demand, respectively. The first number under each bar denotes the market number, and the second denotes the retailer assigned to that market. In this research, each unit of cost is equivalent to 1000 dollars. As shown in Fig. 3.1, the highest amount of fulfilled demand is 6483,

Table 3.2 Results of chosen options in the product 1

Product 1														
	Mandatory module											Arbitrary module		
	MDLE-1	MDLE-2	MDLE-3	MDLE-4	MDLE-5	MDLE-6	MDLE-7	MDLE-8	MDLE-9	MDLE-10	MDLE-11	MDLE-12	MDLE-13	MDLE-14
Modules	LCD	CPU	Motherboard	RAM	Chipset	PCtel modem	Hard drive	Case	Battery	GPU	Graphic cash	Memory card	Blue ray	Bluetooth
Options	OPTN-1	OPTN-1	OPTN-1	OPTN-1	OPTN-1	OPTN-1	OPTN-1	OPTN-1	OPTN-1	OPTN-1	OPTN-1	OPTN-1	OPTN-1	OPTN-1
	OPTN-2	OPTN-2	OPTN-2	OPTN-2	OPTN-2	OPTN-2	OPTN-2	OPTN-2	OPTN-2	OPTN-2	OPTN-2	OPTN-2	OPTN-2	OPTN-2
	OPTN-3	OPTN-3	OPTN-3	OPTN-3	OPTN-3	OPTN-3	OPTN-3	OPTN-3	OPTN-3	OPTN-3	OPTN-3	OPTN-3	OPTN-3	OPTN-3

Table 3.3 Results of chosen options in the product 2

Product 2														
	Mandatory module											Arbitrary module		
	MDLE-1	MDLE-2	MDLE-3	MDLE-4	MDLE-5	MDLE-6	MDLE-7	MDLE-8	MDLE-9	MDLE-10	MDLE-11	MDLE-12	MDLE-13	MDLE-14
Modules	LCD	CPU	Motherboard	RAM	Chipset	PCtel modem	Hard drive	Case	Battery	GPU	Graphic cash	Memory card	Blue ray	Bluetooth
Options	OPTN-1	OPTN-1	OPTN-1	OPTN-1	OPTN-1	OPTN-1	OPTN-1	OPTN-1	OPTN-1	OPTN-1	OPTN-1	OPTN-1	OPTN-1	OPTN-1
	OPTN-2	OPTN-2	OPTN-2	OPTN-2	OPTN-2	OPTN-2	OPTN-2	OPTN-2	OPTN-2	OPTN-2	OPTN-2	OPTN-2	OPTN-2	OPTN-2
	OPTN-3	OPTN-3	OPTN-3	OPTN-3	OPTN-3	OPTN-3	OPTN-3	OPTN-3	OPTN-3	OPTN-3	OPTN-3	OPTN-3	OPTN-3	OPTN-3

Table 3.4 Sensitivity analysis for the increasing quality level of the product 1

Product 1														
Modules	Mandatory module											Arbitrary module		
	MDLE-1	MDLE-2	MDLE-3	MDLE-4	MDLE-5	MDLE-6	MDLE-7	MDLE-8	MDLE-9	MDLE-10	MDLE-11	MDLE-12	MDLE-13	MDLE-14
	LCD	CPU	Motherboard	RAM	Chipset	PCtel modem	Hard drive	Case	Battery	GPU	Graphic cash	Memory card	Blue ray	Bluetooth
Options	OPTN-1	OPTN-1	OPTN-1	OPTN-1	OPTN-1	OPTN-1	OPTN-1	OPTN-1	OPTN-1	OPTN-1	OPTN-1	OPTN-1	OPTN-1	OPTN-1
	OPTN-2	OPTN-2	OPTN-2	OPTN-2	OPTN-2	OPTN-2	OPTN-2	OPTN-2	OPTN-2	OPTN-2	OPTN-2	OPTN-2	OPTN-2	OPTN-2
	OPTN-3	OPTN-3	OPTN-3	OPTN-3	OPTN-3	OPTN-3	OPTN-3	OPTN-3	OPTN-3	OPTN-3	OPTN-3	OPTN-3	OPTN-3	OPTN-3

Table 3.5 Sensitivity analysis for the increasing price of the product 2

Product 2														
	Mandatory module											Arbitrary module		
	MDLE-1	MDLE-2	MDLE-3	MDLE-4	MDLE-5	MDLE-6	MDLE-7	MDLE-8	MDLE-9	MDLE-10	MDLE-11	MDLE-12	MDLE-13	MDLE-14
Modules	LCD	CPU	Motherboard	RAM	Chipset	PCtel modem	Hard drive	Case	Battery	GPU	Graphic cash	Memory card	Blue ray	Bluetooth
Options	OPTN-1	OPTN-1	OPTN-1	OPTN-1	OPTN-1	OPTN-1	OPTN-1	OPTN-1	OPTN-1	OPTN-1	OPTN-1	OPTN-1	OPTN-1	OPTN-1
	OPTN-2	OPTN-2	OPTN-2	OPTN-2	OPTN-2	OPTN-2	OPTN-2	OPTN-2	OPTN-2	OPTN-2	OPTN-2	OPTN-2	OPTN-2	OPTN-2
	OPTN-3	OPTN-3	OPTN-3	OPTN-3	OPTN-3	OPTN-3	OPTN-3	OPTN-3	OPTN-3	OPTN-3	OPTN-3	OPTN-3	OPTN-3	OPTN-3

Table 3.6 Sensitivity analysis for decreasing fixed development cost of the product 2

Product 2														
	Mandatory module											Arbitrary module		
	MDLE-1	MDLE-2	MDLE-3	MDLE-4	MDLE-5	MDLE-6	MDLE-7	MDLE-8	MDLE-9	MDLE-10	MDLE-11	MDLE-12	MDLE-13	MDLE-14
Modules	LCD	CPU	Motherboard	RAM	Chipset	PCtel modem	Hard drive	Case	Battery	GPU	Graphic cash	Memory card	Blue ray	Bluetooth
Options	OPTN-1	OPTN-1	OPTN-1	OPTN-1	OPTN-1	OPTN-1	OPTN-1	OPTN-1	OPTN-1	OPTN-1	OPTN-1	OPTN-1	OPTN-1	OPTN-1
	OPTN-2	OPTN-2	OPTN-2	OPTN-2	OPTN-2	OPTN-2	OPTN-2	OPTN-2	OPTN-2	OPTN-2	OPTN-2	OPTN-2	OPTN-2	OPTN-2
	OPTN-3	OPTN-3	OPTN-3	OPTN-3	OPTN-3	OPTN-3	OPTN-3	OPTN-3	OPTN-3	OPTN-3	OPTN-3	OPTN-3	OPTN-3	OPTN-3

Table 3.7 Sensitivity analysis for increasing production cost of the product 2

Product 2														
	Mandatory module											Arbitrary module		
	MDLE-1	MDLE-2	MDLE-3	MDLE-4	MDLE-5	MDLE-6	MDLE-7	MDLE-8	MDLE-9	MDLE-10	MDLE-11	MDLE-12	MDLE-13	MDLE-14
Modules	LCD	CPU	Motherboard	RAM	Chipset	PCtel modem	Hard drive	Case	Battery	GPU	Graphic cash	Memory card	Blue ray	Bluetooth
Options	OPTN-1	OPTN-1	OPTN-1	OPTN-1	OPTN-1	OPTN-1	OPTN-1	OPTN-1	OPTN-1	OPTN-1	OPTN-1	OPTN-1	OPTN-1	OPTN-1
	OPTN-2	OPTN-2	OPTN-2	OPTN-2	OPTN-2	OPTN-2	OPTN-2	OPTN-2	OPTN-2	OPTN-2	OPTN-2	OPTN-2	OPTN-2	OPTN-2
	OPTN-3	OPTN-3	OPTN-3	OPTN-3	OPTN-3	OPTN-3	OPTN-3	OPTN-3	OPTN-3	OPTN-3	OPTN-3	OPTN-3	OPTN-3	OPTN-3

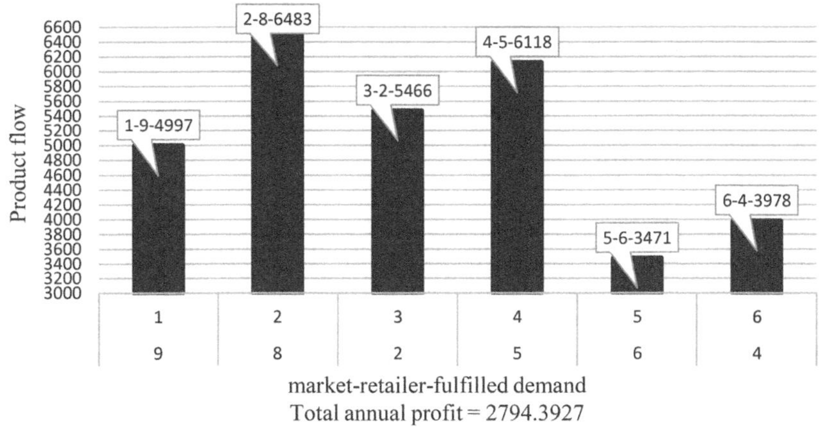

Fig. 3.1 The solution of the model for optimal locations

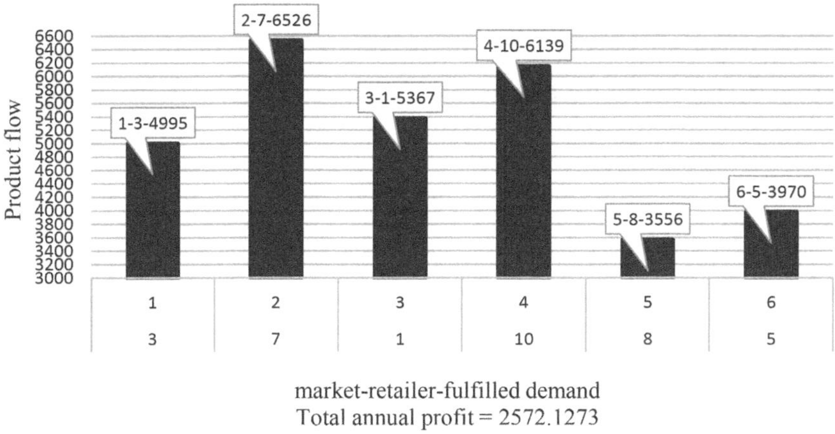

Fig. 3.2 The solution of the model for random locations

related to Market 2 and Retailer 8. The lowest amount of fulfilled demand is 3471, related to market 5 and retailer 6. Also, the total annual profit is obtained at 2794.39.

To examine the effect of changing retailers' locations on the solution, we have developed a sensitivity analysis by assigning fixed random locations for retailers. Figure 3.2 displays the solution of the proposed model for the first scenario considering retailers' random locations. As shown in Fig. 3.2, the highest amount of fulfilled demand is 6526, related to Market 2 and Retailer 7. The lowest amount of fulfilled demand is 3556, which is related to market 5 and retailer 8. Also, the total annual profit is obtained at 2572.13. The results of Figs. 3.1 and 3.2 indicate that the order of the fulfilled demands from the largest to the smallest does not change with the change in the location of the retailers, but the total annual profit reduces from 2794.39 to 2572.13.

3.4.2 *Dimensional Analysis*

This section develops a dimensional analysis for nine different test problems. These problems are divided into three classes. The first class constitutes a small class with three test problems (P1–P3). The second one constitutes a medium class with three test problems (P4–P6). The last one constitutes a large class with three test problems (P7–P9). The dimensional analysis of the model is done by using GAMS software [31] for exact solutions and MATLAB R2019b software [32] for GASA solutions on a computer with Intel(R) Core i5 2450M CPU and 2.5 GHz processor with 6.00 GB RAM. Table 3.8 shows the configuration of the test problems. The quantity of discrete variables, total variables, constraints, and the output of CPU time obtained after solving each test problem are presented in Table 3.9.

Table 3.8 Configuration of the test problems

Class	Problem	Quantity of modules	Quantity of options	Quantity of markets	Quantity of retailers	Quantity of scenarios
Small	P1	12	2	3	8	5
	P2	13	2	4	9	10
	P3	13	3	5	9	13
Medium	P4	14	3	6	10	15
	P5	15	3	6	11	16
	P6	15	4	6	11	16
Large	P7	16	4	7	12	18
	P8	16	4	7	12	20
	P9	17	4	8	12	20

Table 3.9 Calculating results of test problems

Class	Problem	Quantity of discrete variables	Total quantity of variables	Quantity of constraints	CPU time (second)
Small	P1	342	385	1818	3.237
	P2	840	943	3060	12.688
	P3	1338	1497	5765	36.367
Medium	P4	1992	2205	7585	104.376
	P5	2319	2546	8762	450.039
	P6	2364	2591	11972	1489.723
Large	P7	3306	3597	14876	2355.031
	P8	3642	3965	15584	2813.866
	P9	4146	4509	17659	3155.257

According to Tables 3.8 and 3.9, it can be concluded that the simultaneous growth in the quantity of modules, the quantity of markets, the quantity of retailers, and the quantity of scenarios leads to a significant enhancement in the quantity of constraints, the total quantity of variables, and the CPU time. The simultaneous growth in the quantity of modules, the quantity of retailers, and the quantity of scenarios enhances the quantity of constraints and the CPU time. Finally, the growth in the quantity of scenarios only leads to increasing the CPU time, significantly.

It is needed to determine the complexity of the model by investigating the CPU time growth trend. For this purpose, we examined the CPU time of the exact solutions with GAMS versus the total quantity of variables and the quantity of constraints in Figs. 3.3 and 3.4, respectively. The results indicate that the CPU times of the test problems grow sharply; therefore, the proposed model is NP-hard. Consequently, an evolutionary algorithm must be developed to solve the model to gain near-optimal responses with a reasonable CPU time.

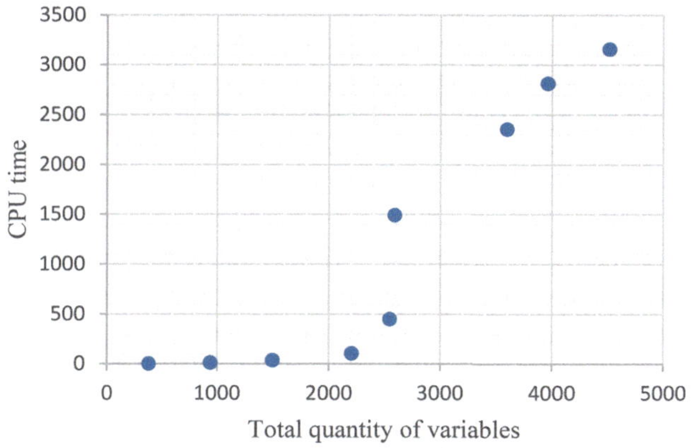

Fig. 3.3 The CPU time growth trend of the GAMS solution versus the total quantity of variables

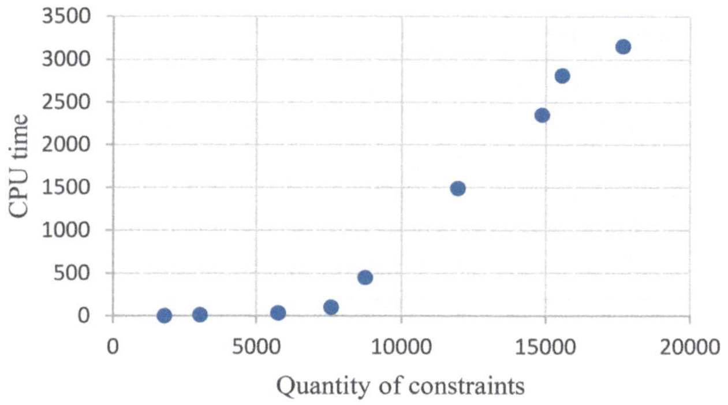

Fig. 3.4 The CPU time growth trend of the GAMS solution versus the quantity of constraints

3.4.3 *Hybrid Genetic Algorithm and Simulated Annealing Approach*

As mentioned in Sect. 3.4.2, the development of an evolutionary algorithm is required for NP-hard problems, like the proposed model in this research, to gain near-optimal responses with a reasonable CPU time. Numerous evolutionary algorithms have already been presented to solve product family problems. In the meantime, some researchers such as Jiao et al. [33], Huang et al. [34], and Khajavirad et al. [35] have concluded that the genetic algorithm is effective for solving such problems. According to Hejazi et al. [36], the GASA has a higher convergence speed than the genetic algorithm. Hence, we apply GASA to solve the proposed model.

In the GASA procedure, first, an initial solution is produced by the genetic algorithm so that an initial population is randomly selected from the solution space. Next, the roulette wheel selection method selects parents from this population. Afterward, the crossover and mutation are applied to the parents. In this research, we developed the arithmetic crossover, which linearly combines the two parent chromosomes. Also, we developed the real mutation by adding a normal random value to the old value of the solution. Then the offspring are evaluated, and to decide about selecting parents and offspring, the simulated annealing method will be used. For this purpose, the fitness function of the offspring is compared with that of the parents. If the offspring's fitness function is better than the parents', that offspring is accepted; otherwise, it is accepted if Boltzmann's probability function is met. This probability function is calculated through $\delta \leq e^{(\frac{-\Delta E}{T})}$ in which ΔE is the difference between the fitness function of the offspring and the parents, T denotes the temperature, and δ indicates a value that is uniformly distributed between 0 and 1. Finally, if the termination conditions are not fulfilled, the temperature is reduced, and the next iteration is executed [37–39].

3.4.4 *Parameter Tuning*

In this chapter, the parameters of the GASA are adjusted by Taguchi method. This method optimizes the process design problems cost-effectively by decreasing process variability through the robust design of experiments. Taguchi method analyzes factors using orthogonal arrays. An orthogonal array designs some experiments considering the number of levels and factors, and then it determines the appropriate level of each factor according to the analysis of experiments. Also, to reach optimal level of parameters, the signal-to-noise ratio needs to be maximized. The signal-to-noise ratio optimizes the mean and variance of the response simultaneously. Taguchi

Table 3.10 The specified levels of factors for GASA algorithm

	Population size	Generation	Sub-iteration	Crossover	Mutation	Initial temperature
Level 1	70	150	10	0.8	0.2	10
Level 2	100	200	20	1	0.4	50

signal-to-noise ratio for maximization problems is calculated through the following equation [19, 39]:

$$S/N = -10\log_{10}\left[\frac{1}{n}\sum_{j=1}^{n}\left(\frac{1}{y_j}\right)^2\right], \tag{3.19}$$

where y_j denotes the response value for the yth experiment and n represents the replication number. In this chapter, six factors and two levels of parameters are considered. These factors include population size, generation, sub-iteration, crossover, mutation, and initial temperature. As a result, Taguchi method L8 (2^6 factorial design) has been applied for test problem P9, which is the largest size of the test problems. Also, for more accuracy, the model has been replicated three times. The levels of the factors including level 1 and level 2 have been explained in Table 3.10. The comprehensive information of these experiments and their S/N ratios for test problem P9 are given in Table 3.11. Figures 3.5 and 3.6 represent the Taguchi design analysis for means and S/N ratios, respectively. The fitness function S/N ratio has been considered as "larger is better." Therefore, the larger the mean and the S/N ratio, the better. Consequently, the best level of factors for the GASA algorithm is considered in Table 3.12.

3.4.5 Results

The analysis of the results for the GASA algorithm and the exact solutions with GAMS will be presented in this section. As mentioned before, we have employed the GASA to solve the model because of the NP-hard construction of the proposed model and the efficiency of the GASA to solve similar problems. So, considering the best levels of the GASA parameters which are illustrated in Table 3.12, the model has been executed by the GASA three times to reach reliable solutions. The CPU time and function outputs of the GASA and the GAMS solutions for problems P1–P9 are explained in Table 3.13. According to Table 3.13, the small standard deviation between the three replications of the GASA fitness functions demonstrates that more replications are not needed. Figure 3.7 shows the convergence plot of the GASA algorithm for test problem P9. The GASA process in Fig. 3.7 starts from 3100 for the fitness function and converges to its maximum value at 3526.

Figure 3.8 depicts the comparison between the CPU times of the GAMS solutions and the GASA algorithm. As shown in Fig. 3.8, the CPU times of the GASA

Table 3.11 The designed experiments and S/N ratios

Number of experiments	Population size	Generation	Sub-iteration	Crossover	Mutation	Initial temperature	Fitness S/N ratio
1	70	150	10	0.8	0.2	10	70.141
2	70	150	10	1	0.4	50	70.159
3	70	200	20	0.8	0.2	50	70.177
4	70	200	20	1	0.4	10	70.21
5	100	150	20	0.8	0.4	10	70.808
6	100	150	20	1	0.2	50	70.84
7	100	200	10	0.8	0.4	50	70.896
8	100	200	10	1	0.2	10	71.101

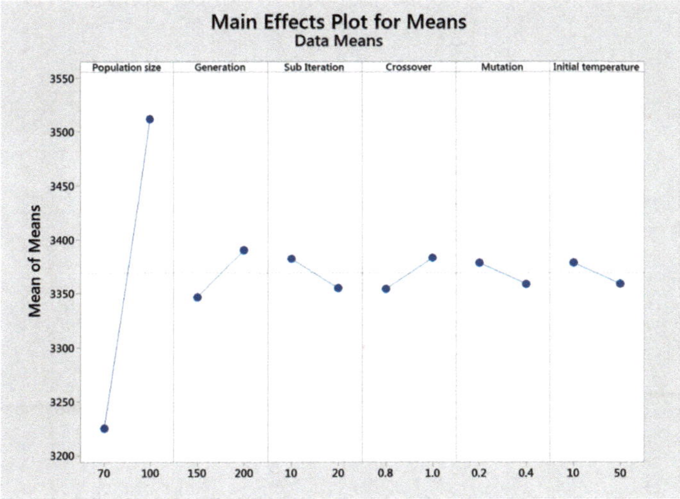

Fig. 3.5 The Taguchi design means plot for GASA

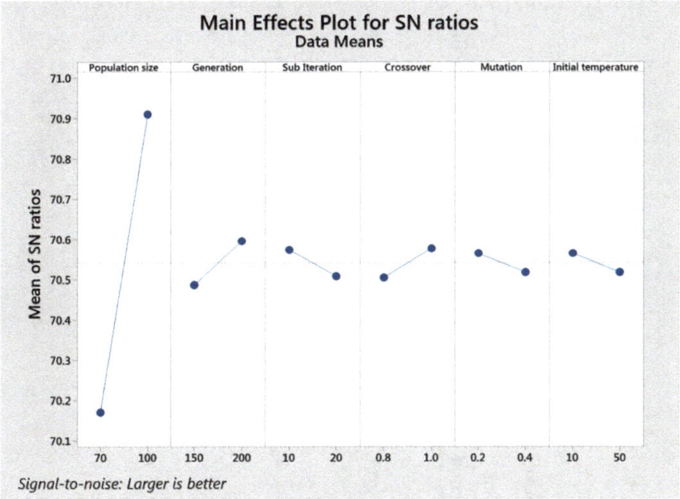

Fig. 3.6 The Taguchi design S/N ratio plot for GASA

Table 3.12 The best level of factors for the GASA algorithm

Population size	Generation	Sub-iteration	Crossover	Mutation	Initial temperature
100	200	10	1	0.2	10

Table 3.13 The outputs of the GAMS and the GASA

	GAMS solutions		GASA solutions		
Problem number	Objective function	CPU time (second)	Average fitness function	Average CPU time (second)	Fitness function standard deviation
1	1550.5228	3.237	1995.968	190.886	8.19
2	2119.9124	12.688	2256.134	251.991	12.524
3	2324.6878	36.367	2390.606	264.444	2.989
4	2777.6279	104.376	2863.842	296.746	14.157
5	2778.9423	450.039	2907.273	317.64	6.253
6	2849.723	1489.723	2992.6	322.534	5.344
7	3196.9678	2355.031	3181.548	342.679	2.565
8	3200.0059	2813.866	3341.354	376.37	14.029
9	3812.645	3155.257	3589.856	377.779	7.989

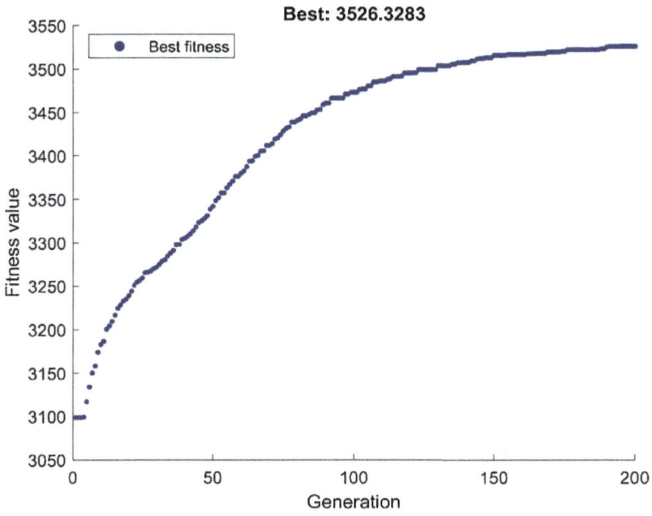

Fig. 3.7 The GASA plot for problem P9

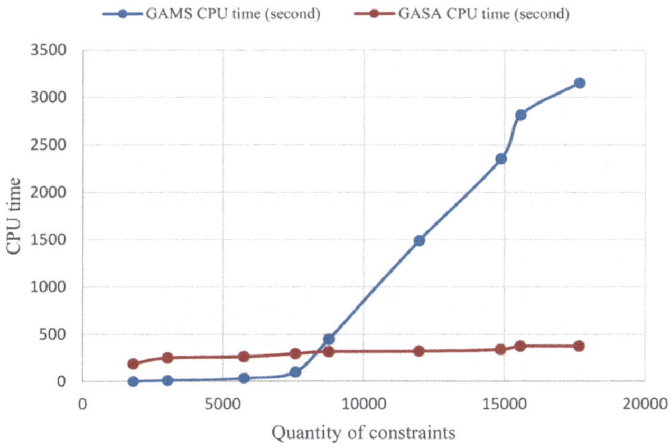

Fig. 3.8 The comparison between the average CPU times of the GAMS and GASA solutions

algorithm are more than those of the GAMS for test problems P1–P4 due to the iterative nature of the GASA algorithm. For test problems P5–P9, it can be observed that the GASA has better CPU time than the GAMS. Also, the CPU times of the GAMS increase sharply, while those of the GASA grow gradually in a linear pattern. As a result, using the GASA algorithm leads to significant time reduction in solving the proposed model. Figure 3.9 shows the comparison chart of the average objective functions of the GAMS and GASA solutions for test problems P1–P9. Overall, it is apparent that the results obtained by the GASA are very close to those of the GAMS solutions. Also, it can be observed that in all test problems except P7 and P9, the

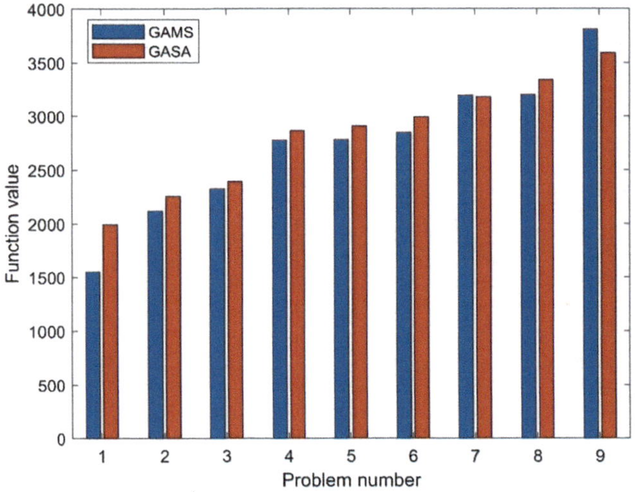

Fig. 3.9 The comparison between the average objective functions of the GAMS and GASA solutions

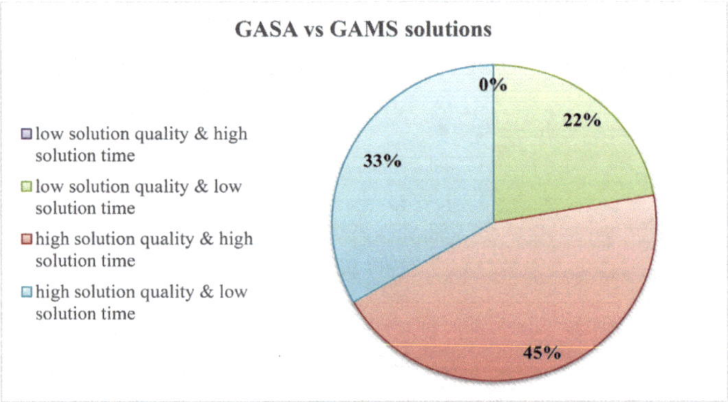

Fig. 3.10 The performance results of the GASA versus GAMS

GASA has better performance than GAMS solutions. To measure and provide more details about the performance of the GASA algorithm, the following relations are employed [40]:

The comparison between the performance of the GASA algorithm and GAMS solutions according to the above equations is represented in Fig. 3.10. According to Fig. 3.10, in 22% of the test problems (test problems P7 and P9), the GASA ameliorates the solution time but causes worse objective function quality. In 45% of the test problems (test problems P1–P4), the GASA gains a better objective function quality, but it is more time-consuming. In 33% of the test problems (test problems P5, P6, and P8), the GASA performed better than GAMS solutions in terms of both the solution time and the quality of the objective function. Eventually, there is no test problem (0%) with the GASA performing worse than the GAMS in respect of the time and the objective function quality. As a result, the GASA algorithm could save an average of 853.28 seconds in solution time and improve an average of 100.91 units in the objective function quality.

3.5 Conclusion

In this chapter, we studied a simultaneous supply chain and product family design problem in a stochastic three-echelon supply chain, including a manufacturer, retailers, and markets under demand uncertainty. In the product design process, we considered two types of modules (mandatory and arbitrary) for each product and two options (mandatory and arbitrary) for each module which the customer could choose among arbitrary modules and options. Consequently, various products with different costs and quality levels were generated. Therefore, in the proposed model, two types of customers were considered. One group preferred cheaper products, and the others preferred higher quality products. The objective function of the proposed model was defined as profit maximization in which the costs of fixed development, transportation, and retailer locating were regarded. In addition, the locating and allocating of retailers to the markets were considered in this research. Regarding solution methods, we solved the model with the exact method, analyzed the results by sensitivity analysis, and then developed a dimensional analysis to evaluate the behavior of the model in large dimensions. The results indicated that the model is NP-hard. Thus, we developed the GASA to gain near-optimal responses with a reasonable CPU time. Finally, a comparison between the exact solutions and the GASA results was made, demonstrating the GASA's superiority over the exact method. For future studies, we recommend incorporating the efficiency of all supply chain elements into the proposed model, as well as integrating a reverse supply chain with circularity considerations, including green transportation.

Appendix

The proposed model data is considered as follows:

$A_{ss}^{k} = triangular(3096, 4946.5, 6392)$
$E_n = ceil(uniform(500, 600))$
$mp_n = 0.15$
$qq_{cu} = uniform(10, 12.5)$
$qM_m^j = uniform(0.7, 1)$
$qF_f^j = uniform(0.6, 1)$
$CM_m^j = uniform(0.08, 0.12) * qM_m^j$

$$CF_f^j = uniform(0.03, 0.05) * qF_f^j$$
$$DM_m^j = ceil(uniform(25, 40) * (qM_m^j))$$
$$DF_f^j = ceil(uniform(10, 15) * (qF_f^j))$$
$$Sc_n^k = uniform(1, 3)$$
$$cc^r = uniform(0.04, 0.06)$$
$$L_r^k = uniform(3, 4)$$
$$B_1^k = 0.45$$
$$B_2^k = 0.55$$
$$rr_1^k = 20$$
$$rr_2^k = 1$$
$$G_1^k = 3$$
$$G_2^k = 30$$
$$pi_{ss} = \frac{1}{ss}$$

References

1. Simpson, T.W., Siddique, Z., Jiao, R.J.: Product platform and product family design: methods and applications. Springer Science & Business Media (2006).
2. Rezapour, S., Hassani, A., Farahani, R. Z.: Concurrent design of product family and supply chain network considering quality and price. Transp. Res. Part E: Logist. Transp. Rev. **81**, 18–35 (2015).
3. Mirzapour Al-E-Hashem, S.M.J., Malekly, H., Aryanezhad, M.B.: A multi-objective robust optimization model for multi-product multi-site aggregate production planning in a supply chain under uncertainty. Int. J. Prod. Econ. **134**(1), 28–42 (2011).
4. Zeidabadi, S.A., Rezaee Nik, E., Hejazi, T.-H.: Multi-objective robust production planning considering workforce efficiency with a metaheuristic solution approach. *Int. J. Ind. Eng.* **30**(1) (2023).
5. Agard, B., Penz, B.: A simulated annealing method based on a clustering approach to determine bills of materials for a large product family. Int. J. Prod. Econ. **117**, 389–401 (-)
6. Briant, O., Naddef, D.: The optimal diversity management problem. Oper. Res. **52**(4), 515–526 (2004).
7. Wu, L., De Matta, R., Lowe, T.J.: Updating a modular product: How to set time to market and component quality. IEEE Trans. Eng. Manag. **56**(2), 298–311 (2009).
8. Das, K., Chowdhury, A.H.: Designing a reverse logistics network for optimal collection, recovery and quality-based product-mix planning. Int. J. Prod. Econ. **135**(1), 209–221 (2012).
9. Ma, J., Kremer, G.E.O.: A sustainable modular product design approach with key components and uncertain end-of-life strategy consideration. Int. J. Adv. Manuf. Technol. **85**, 741–763 (2016).
10. Olivares Aguila, J., ElMaraghy, W., ElMaraghy, H.: Impact of risk attitudes on the concurrent design of supply chains and product architectures. Procedia CIRP **81**, 974–979 (2019)
11. Gan, T.-S., Steffan, M., Grunow, M., Akkerman, R.: Concurrent design of product and supply chain architectures for modularity and flexibility: process, methods, and application. Int. J. Prod. Res. **60**(7), 2292–2311 (2022).
12. Ziaei, M., Ketabi, S., Ghandehari, M.: Concurrent optimizing configuration, price, warranty, and supply policy in configurable product family (CPF). *J. Eng. Des.* **33**(1), 15–38 (2022).

13. Hossain, M.S., Chakrabortty, R.K., El Sawah, S., Ryan, M.J.: A multi-objective Bi-level leader-follower joint optimization for concurrent design of product family and assembly system. Comput. Ind. Eng. **177**, 109035 (2023).
14. Hossain, M.S., Chakrabortty, R.K., Elsawah, S., Ryan, M.J.: Modelling and application of hierarchical joint optimisation for modular product family and supply chain architecture. Int. J. Adv. Manuf. Technol. **126**(3), 947–971 (2023).
15. Luo, X., Du, Y., Chen, Q., Zhang, Z.: Joint optimization of low-carbon product family configuration and smart production line selection. Comput. Ind. Eng. **182**, 109403 (2023).
16. Franca, R.B., Jones, E.C., Richards, C.N., Carlson, J.P.: Multi-objective stochastic supply chain modeling to evaluate tradeoffs between profit and quality. Int. J. Prod. Econ. **127**(2), 292–299 (2010).
17. Yin, S., Nishi, T.: A supply chain planning model with supplier selection under uncertain demands and asymmetric information. Procedia CIRP **17**, 639–644 (2014).
18. Nagurney, A., Saberi, S., Shukla, S., Floden, J.: Supply chain network competition in price and quality with multiple manufacturers and freight service providers. Transp. Res. Part E: Logist. Transp. Rev. **77**, 248–267 (2015).
19. Hejazi, T.H.: A scenario-based desirability function for correlated multi-response optimization problems considering modeling and implementation errors. J. Comput. Sci. **63**, 101764 (2022).
20. Hejazi, T.-H., Khorshidvand, B.: Robust optimization of sustainable closed-loop supply chain network considering product family. Environ. Dev. Sustain. **26**(4), 10591–10621 (2024).
21. Mridha, B., Pareek, S., Goswami, A., Sarkar, B.: Joint effects of production quality improvement of biofuel and carbon emissions towards a smart sustainable supply chain management. J. Clean. Prod. **386**, 135629 (2023).
22. Huma, S., Ahmed, W., Zaman, S.U.: The impact of supply chain quality integration on a firm's sustainable performance. TQM J. **36**(2), 385–404 (2023).
23. Baud-Lavigne, B., Bassetto, S., Agard, B.: A method for a robust optimization of joint product and supply chain design. J. Intell. Manuf. **27**, 741–749 (2016)
24. Hejazi, T.-H., Seyyed-Esfahani, M.: Multistage-multiresponse models for dynamic quality chain design problems. Qual. Reliab. Eng. Int. **33**(6), 1263–1279 (2017).
25. Homayounmehr, A., Hejazi, T.-H.: Optimal design of a multi-period supply chain network with unreliable elements: A two-stage stochastic programming approach. J. Ind. Eng. Manag. Stud. **9**(1), 120–135 (2022).
26. G'omez-Rocha, J.E., Hern'andez-Gress, E.S.: A stochastic programming model for multi-product aggregate production planning using valid inequalities. Appl. Sci. **12**(19), 9903 (2022).
27. Foroozesh, N., Karimi, B., Mousavi, S.M.: Green-resilient supply chain network design for perishable products considering route risk and horizontal collaboration under robust interval-valued type-2 fuzzy uncertainty: A case study in food industry. J. Environ. Manag. **307**, 114470 (2022).
28. Zarouk, Y., Mahdavi, I., Rezaeian, J., Santos-Arteaga, F.J.: A novel multi-objective green vehicle routing and scheduling model with stochastic demand, supply, and variable travel times. *Comput. Oper. Res.* **141**, 105698 (2022).
29. Wang, Q., Li, H., Wang, D., Cheng, T.C.E., Yin, Y.: Bi-objective perishable product delivery routing problem with stochastic demand. Comput. Ind. Eng. **175**, 108837 (2023).
30. Bayram, V., Yıldız, B., Farham, M.S.: Hub network design problem with capacity, congestion, and stochastic demand considerations. Transp. Sci. **57**(5), 1276–1295 (2023)
31. GAMS Development Corporation: General Algebraic Model System (GAMS 24.7.3). GAMS Development Corporation, Fairfax, VA (2016) Available at: https://www.gams.com.
32. The MathWorks Inc.: MATLAB version: R2019b (No. R2019b). The MathWorks Inc. (2019) Available at: https://www.mathworks.com.
33. Jiao, J., Zhang, Y., Wang, Y.: A generic genetic algorithm for product family design. J. Intell. Manuf. **18**, 233–247 (2007).
34. Huang, G.Q., Li, L., Schulze, L.: Genetic algorithm-based optimisation method for product family design with multi-level commonality. J. Eng. Des. **19**(5), 401–416 (2008).

35. Khajavirad, A., Michalek, J.J., Simpson, T.W.: An efficient decomposed multiobjective genetic algorithm for solving the joint product platform selection and product family design problem with generalized commonality. Struct. Multidisc. Optim. **39**, 187–201 (2009).
36. Hejazi, T.H., Asadi Zeidabadi, S., Abbasi, M.: Quality-based design of supply chains considering supplier and technology effects. Soft Comput. **26**(12), 5741–5763 (2022).
37. Metropolis, N., Rosenbluth, A.W., Rosenbluth, M.N., Teller, A.H., Teller, E.: Equation of state calculations by fast computing machines. J. Chem. Phys. **21**(6), 1087–1092 (1953).
38. "Orkc"u, H. H.: Subset selection in multiple linear regression models: A hybrid of genetic and simulated annealing algorithms. Appl. Math. Comput. **219**(23), 11018–11028 (2013).
39. Torkaman, S., Ghomi, S.M.T.F., Karimi, B.: Hybrid simulated annealing and genetic approach for solving a multi-stage production planning with sequence-dependent setups in a closed-loop supply chain. Appl. Soft Comput. **71**, 1085–1104 (2018).
40. Badri, H., Bashiri, M., Hejazi, T.H.: Integrated strategic and tactical planning in a supply chain network design with a heuristic solution method. Comput. Oper. Res. **40**, 1143–1154 (2013)

Chapter 4
Advancing Sustainability Through Operational Research in Circular Supply Chains: Trends and Opportunities

Habib Chabchoub, Arij Lahmar, and Areej Aftab Siddiqui

Abstract The synergy between operational research (OR) and circular supply chain management (CSCM) has gained significant attention in recent years as businesses strive to minimize waste and resource use while optimizing decision-making and efficiency. This chapter presents a comprehensive review of the literature at the intersection of OR and CSCM, highlighting the applications of OR techniques in closed-loop supply chains, green logistics, and waste minimization. The study identifies key themes, gaps in existing research, and opportunities for advancing the integration of OR and CSCM. A structured literature search and analysis of 402 articles from 84 academic journals was conducted. Descriptive analysis reveals a growing trend in publications since the early 2000s, with a sharp rise around 2020. Content analysis using co-occurrence of keywords identifies four main clusters: operational efficiency and cost management, sustainability and environmental management, advanced optimization techniques, and exact optimization methods. The temporal evolution of clusters shows a shift from traditional supply chain optimization to the incorporation of circular economy concepts and sustainability. The discussion highlights the role of traditional OR methods and the potential of advanced techniques such as machine learning, artificial intelligence, and cutting-plane algorithms in addressing the increasing complexity of CSCM. Limitations and future research directions are discussed, emphasizing the need for more rigorous evidence-based research, multi-objective programming, and the development of standardized data frameworks for the effective application of AI and ML in CSCM.

Keywords Sustainability · Supply chain · Operational research · Circular economy · Waste · Recycling · Decision · Optimization

H. Chabchoub
Al Ain University, Abu Dhabi, UAE
e-mail: habib.chabchoub@aau.ac.ae

A. Lahmar (✉) · A. A. Siddiqui
Dubai Business School, University of Dubai, Dubai, UAE

B. Jarboui et al. (eds.), *Circular Economy Supply Chains*,
Computational Intelligence Methods and Applications,
https://doi.org/10.1007/978-981-96-7905-8_4

4.1 Introduction

Operational research encompasses a comprehensive and rigorous scientific investigation with the primary objective of devising efficient and effective techniques to effortlessly and adeptly address various operational issues that arise in industrial and economic systems, encompassing a wide range of institutions and organizations, such as businesses, engineering firms, government agencies, and nonprofit organizations [15].

Through an informed utilization of the most advanced and cutting-edge mathematical and computational tools, operational research diligently and assiduously aims to offer practical and tangible solutions to the intricate and multifaceted problems that manifest within these diverse domains. This highly esteemed and dynamic field relies significantly on continual progress and breakthroughs in the realm of mathematical and computational sciences to effectively address and overcome an ever-expanding and diverse array of complex real-world challenges [25]. The profound influence of operational research extends far beyond its practical applications, as it significantly impacts and enriches fundamental theoretical fields such as transportation systems, communication networks, production processes, cost theory, and decision-making theory [28]. By placing a primary emphasis on innovation and forward-thinking approaches, operational research consistently and continuously enhances and implements systematic and methodical approaches to thoroughly scrutinize and critically evaluate decision-making processes to revolutionize and transform the way in which we adeptly and skillfully navigate through complex and ever-evolving environments that are characterized by their intricacy and dynamism [2].

This field has driven the optimization of processes across transport, logistics, production, and decision-making. Traditional approaches focus on efficiency improvement, cost reduction, and the enhancement of operational performance [7]. However, over the last years, there has been significant change in the business landscape. Companies across various industries have become increasingly aware of the environment and importance of sustainability [12]. Managing a circular supply chain requires complete knowledge of the different interlinking components and stakeholders [5, 16]. Businesses and governments must, therefore, make their decisions on economic, environmental, and social impacts. Such a multidimensional perception would therefore ensure that informed choices pertinent to their sustainability objectives are made while considering the long-term feasibility of the operational [17]. Beyond this, advanced means of decision-making should be one of the key widespread adoptions to cope with the complexity of circular supply chains. Such methods may include the use of state-of-the-art technologies, such as data analytics and artificial intelligence, that optimally perform resource allocation, routing, and waste minimization [29]. Therefore, operational research has become a crucial "synergist field" in the context of moving toward a more circular supply chain. As businesses move beyond the linear model, the need for advanced methods to manage material flow, optimize resources, and minimize waste has become necessary. Operational research allows for the modeling and analysis of complex systems in such a way that organizations can evaluate operational, environmental, and financial trade-offs [19].

While the principles of circular economy emphasize resource efficiency, waste reduction, and sustainability, how operational research can offer suitable tools and frameworks to support these objectives on a broader scale remains to be understood. Existing research on this topic is often fragmented and focuses on specific aspects or industries. However, there is still ongoing research on the comprehensive application of operational research in circular supply chains, and there is currently a lack of comprehensive insights in this area. Specific focuses and approaches vary considerably; however, the main gap lies in the absence of a structured approach that can systematically link operational research methodologies to the broader objectives of circular supply chains. The concept of circularity introduces complex issues related to product lifecycle management, resource recovery optimization, and economic-environmental-social trade-offs [17].

The literature does not adequately address how operations research can address these challenges in a way that aligns with the comprehensive vision of the circular economy. Further research is needed to understand how operational research can be adapted to provide integrated solutions for sustainable development across the entire supply chain rather than only addressing isolated applications. In addition, the managerial complexities in balancing the various and often conflicting aims of circular supply chains, such as cost efficiency versus environmental impact, highlight the need for operational research to play a central role in decision-making [19]. This requires a thorough analysis of how operations research models can be expanded to incorporate new variables and trade-offs introduced by circularity without losing sight of broader economic and sustainability objectives [14]. Thus, this study conducts a structured review of the literature, focusing on the use of operational research methods to support decision-making in circular supply chains. Through this analysis, we aim to understand how the operational research academic community has approached the circular economy and identify gaps in research. The remainder of this chapter is organized as follows. Section 4.2 details the research methodology used to locate, identify, and analyze publications. Section 4.3 presents a descriptive analysis of the initial findings, followed by a detailed content analysis in Sect. 4.4. Section 4.5 discusses the implications of this study and future research agendas.

4.2 Research Methodology

The research study was initiated with a comprehensive and thorough literature search to conduct a systematic analysis that provides a comprehensive and detailed overview of existing studies in the field. The authors have considered various publications relevant to this issue. The search involved thorough examination and careful analysis of the titles, abstracts, and keywords of various documents published in

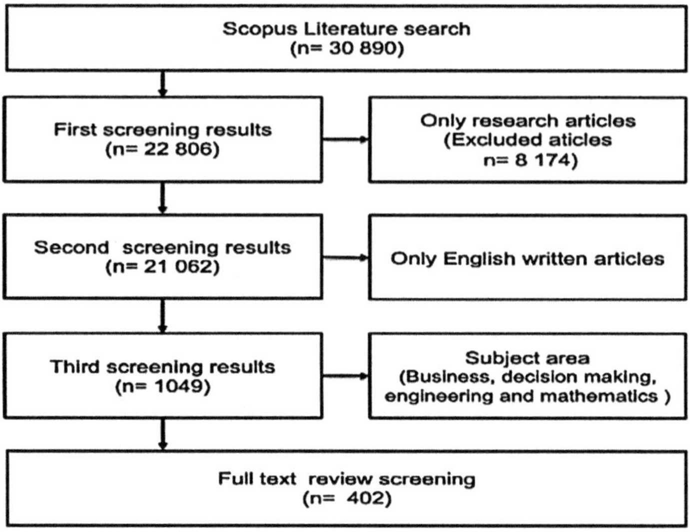

Fig. 4.1 The steps for searching papers for review

reputable scientific journals. The following are a set of well-defined, highly relevant keywords that have been considered by the authors to bring comprehensive results: "operational Research," "Circular Supply Chain," "Circular Supply Chain Model," and "Closed loop Supply Chain." To be able to perform a more detailed search, the authors merged the "Circular Supply Chain" keywords with the subjects that describe the techniques from operational research identified by the organizations EURO and INFORMS: "Simulation," "Optimization," "Queuing Theory," "Markov Decision Processes," "Economic Methods," "Data Analysis," "Statistics," "Neural Networks," "Expert Systems," "Meta-Heuristics," and "Decision Analysis" [3]. In a preliminary search, 30,890 articles were identified and presented to the authors as shown in Fig. 4.1. Only English language journal articles were carefully selected for this comprehensive review. Thus, by taking a fine-tuned and narrowed examination, the authors continued to select only articles that were related to the focus and objectives of this research. The filtering included manual title, abstract, and keyword screening of each paper included in this study to ensure that they were for the purpose and within the scope of the study. Following the above process, 402 carefully selected and studied studies were highly relevant and suitable for the research study. These, in turn, help establish a solid foundation from which the subsequent analysis and conclusions would emanate. To ensure that the selected studies were accurate and relevant, the authors employed restrictions and criteria that ensured that none was subject to biases or discrepancies. This means reevaluating all the keyword and title links emanating from the searches, thereby ensuring the relevance and reliability of the source of information.

4.3 Descriptive Analysis of the Results

The study's descriptive analysis examined trends in publications gathered from academic journals over time. It provides valuable insights into the growing interest in the concept of circularity and the increased utilization of operational research methods in this area, as depicted in Fig. 4.2.

The initial work dates to 1981, with Nahmias [20], an early contributor to the investigation of circularity in inventory management, particularly regarding recycling. His influential study introduced two methods for approximating complex stochastic inventory models. The first involved simplifying the original problem to obtain optimal policies applied to perishable inventory, lost sales due to lead time, and recycling systems. The second method is for continuous review models and utilizes a heuristic approach that incorporates ideas from the Hadley and Whitin model (Q, R), including decaying inventories and estimating the expected backorders in systems with rationing. Subsequently, the number of publications remained minimal, averaging less than one paper per year. This may be attributed to the predominant focus on cost and production efficiency in operational research techniques at the time, as well as the ongoing development of the operational research field. Operational research had an emphasis on traditional models rather than newer, more complex concepts such as recycling and sustainability.

Furthermore, limited computational power and data availability have constrained the comprehensive examination of complex real-world problems. The global emphasis on sustainability and circular economy practices gained traction only in the late 1996s and the early 2004s, which could account for the limited number of publications before this period. This reflects a growing interest in more complex themes within operational research and the gradual but certain incorporation of sustainability and circular economy perspectives in both academic

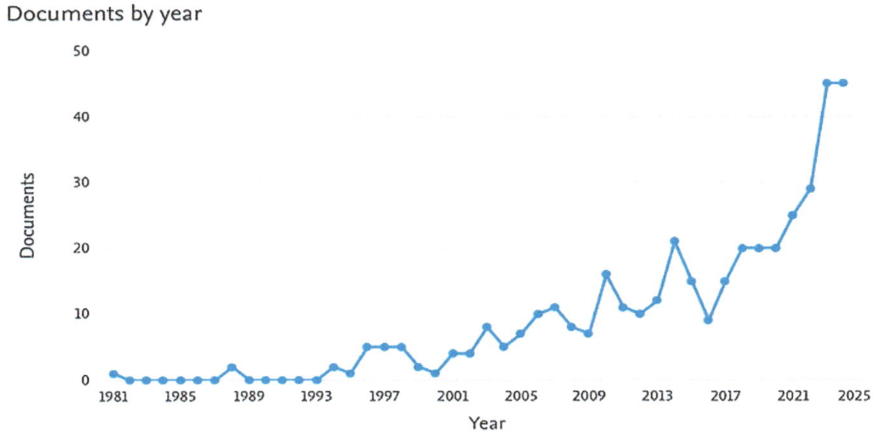

Fig. 4.2 Time evolution of the research publications

and practical discussions. The growth rate was uneven between 2005 and 2010; however, it showed an upward trajectory, indicating consistent expansion. Starting in 2010, the growth of publications was more pronounced, with visible spikes and troughs reflecting increasing interest in these topics. The sharp rise around 2020 reflects that higher numbers of outputs can be observed, owing to improvements in computational technology, data processing, and increased emphasis on sustainability across the globe. In recent years, operational research has begun to take a leading position in resource management, recycling, and eco-development.

The peak number of publications was reached by 2024, signifying a clear pattern of research growth. In other words, interest in operational research, particularly as it pertains to circular economy and sustainability, has been steadily increasing. As a result, this growing interest in this pattern will persist as a crucial and indispensable field of research as the world shifts toward sustainability and seeks new ways to efficiently manage resources [26]. This is evidenced by numerous publications in prominent academic journals such as the European Journal of Operational Research and Annals of Operational Research. A comprehensive analysis identified 402 papers that were published in 84 academic journals. Notably, 36 of these journals featured multiple papers exclusively dedicated to the utilization of operational research tools for circular supply chains. The abundance of academic literature underscores the significance of comprehending and implementing efficient circular supply chain strategies through the lens of operational research (see Fig. 4.3).

The network visualization here presents the coauthorship or citation network in the field of supply chain management, where nodes represent individual researchers, and the edges represent collaborations or citations between them. The size of the nodes likely corresponds to the importance or frequency of collaboration among

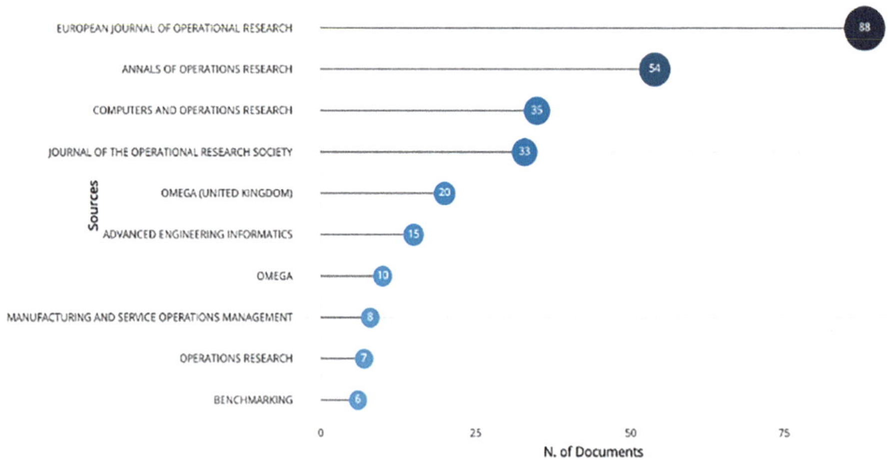

Fig. 4.3 Journals wise distribution of the articles

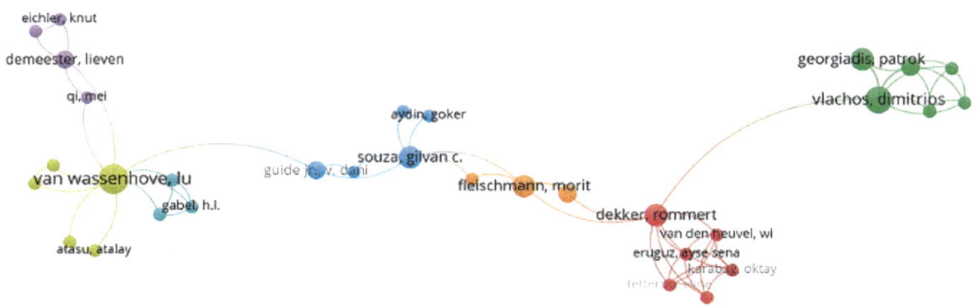

Fig. 4.4 Authorship collaboration network

a certain group of scholars, as does the color of the cluster. The cluster to the left is centered around "Van & Van Wassenhove, L." [27], other hub nodes being "Demeester, Lieven" [9], and "Atasu, Ayalay," a tightly connected research group on sustainability issues or circular supply chains [1]. The central clusters depict strong links between researchers such as "Fleischmann, Moritz" [10] and "Dekker, Rommert" [8], which could point out the influence on the area of reverse logistics or closed-loop supply chains, while the cluster on the right might mirror another school of thought, or research group, in this area, related to green logistics, with "Georgiadis, Patrok" and "Vlachos, Dimitrios" [11].

The network in Fig. 4.4 depicts the links between different authors, highlighting that cross-collaboration exists, but remains limited. The lack of strong links between the clusters can be attributed to the fact that circularity in supply chain research encompasses a wide range of topics. This separation emphasizes the specialized focus of different research groups.

4.4 Content Analysis

Examining the main keywords can be an effective strategy for identifying the primary discussed topics and research themes [4]. Thus, we conducted co-occurrence analysis to identify the most important themes. This method allowed us to gain insights into the specific areas of focus within the research. Using VOSviewer, we input our data and generated a visual representation of the significant terms that appear together. Initially, we identified 2784 important terms, but we established a requirement that they must occur at least seven times to be included in the graph. Ultimately, 93 terms satisfied this criterion and provided a comprehensive overview of the main subjects in the field (Fig. 4.5).

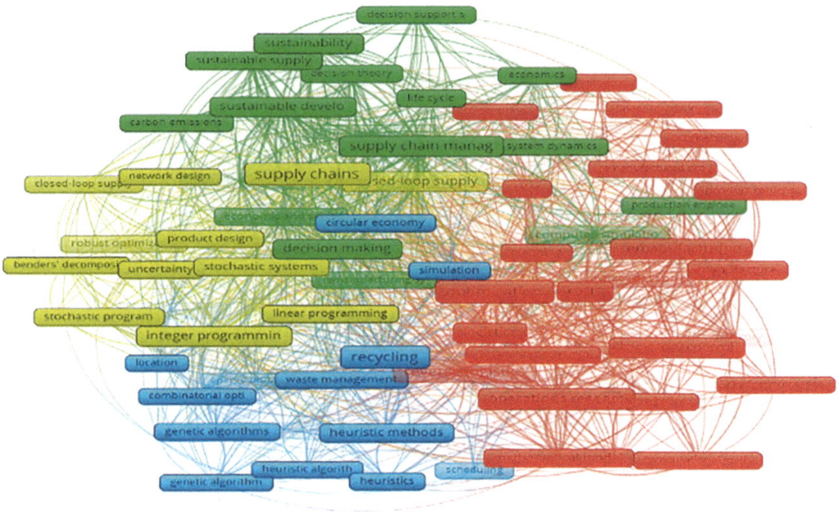

Fig. 4.5 Topics clustering based on keywords co-occurrence analysis

It is remarkable to note from Fig. 4.5 that there are four main clusters in circular supply chain operational, indicating the level of operational research implementation diversity:

- The red cluster deals with the operational efficiency and cost management issues of circular supply chains, with an emphasis on algorithms, dynamic programming, and mathematical modeling. It stresses the main activities of optimization, logistics, and reverse logistics, in keeping with the view that the goals set by sustainability are to be combined with economic performance. The core activities are product remanufacturing, product returns, and inventory control for reverse supply chain processes. This cluster focuses on various classical objectives, which are logistical and cost efficient within product lifecycle management but integrates the implementation of recovery, reuse, and remanufacturing initiatives for resource management and optimization.
- The green cluster is also closely related to circularity regarding the aspects of supply chain sustainability and environmental management, which are integral parts of the circular economy. Keywords such as carbon footprint, environmental impact, and life cycle draw into focus efforts to minimize harm throughout the complete life of a product, from production and disposal to reuse, which is an important focus in circular supply chains. Terms such as decision-making, multi-objective optimization, and system dynamics reflect the relevance of balancing economic, environmental, and social objectives to shift from a traditional linear supply chain to a circular model. Increasingly, remanufacturing systems and sustainable supply chains participate directly in circularity, as the content of these processes is the return of used products to the chain for reuse or recycling, waste

reduction, and resource efficiency enhancement, according to the core principles of the circular economy.

- The blue cluster focuses on technical-advanced optimization techniques with heuristics, genetic algorithms, and combinatorial optimization. It focuses on the operational problems of vehicle routing, scheduling, and resource management for recycling and waste management activities. Research in this cluster has attempted to solve various logistical challenges in relation to the optimal processes of circular supply chains regarding recycling, disassembly, and energy use. These optimization methods are urgently needed to improve the efficiency and sustainability of supply chains that deal with complex material recovery and resource flows.
- The yellow-labeled cluster covers the application of exact methods, which mainly focuses on linear programming, mixed-integer linear programming, and Benders decomposition. Mixed-integer linear programming (MILP) is the most used method for modeling the circular supply chains problems, while Benders decomposition is among the most cited techniques in solving these problems. These approaches were used for a large-scale and complex problem related to closed-loop and reverse supply chains, particularly network design, product design, and robust optimization. The cluster focuses on handling uncertainty and optimizing supply chain processes with both forward and reverse material flows. The aim is to ensure that the flow of circular supply chains is optimal, cost-effective, and sustainable under uncertain variable conditions.

Over the past decade, attention has shifted to the interplay between circularity and operational research tools, as shown in Fig. 4.6. The findings underline that most

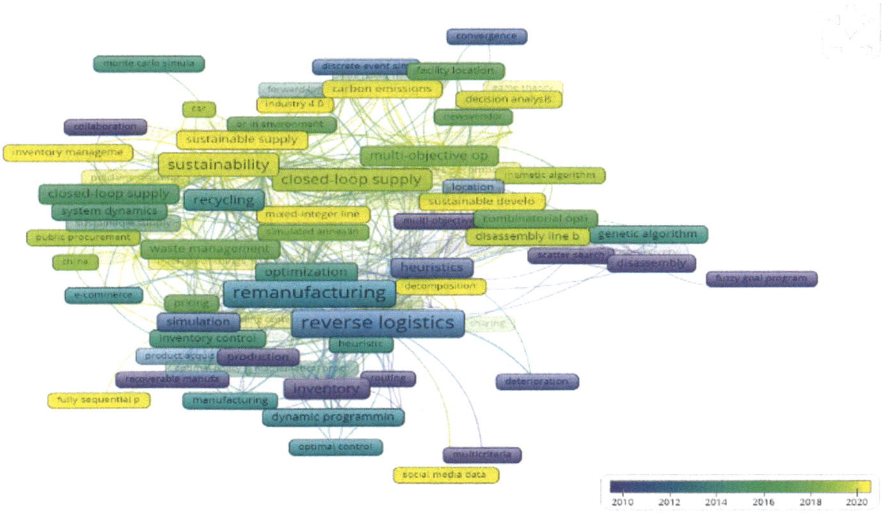

Fig. 4.6 Temporal evolution of the clusters

studies initially focused on traditional supply chain optimization techniques. These tools include linear programming, heuristic methods, and integer programming. Thus, during this period, emphasis was placed on improving supply chain efficiency, logistics, cost management, and inventory control by considering recycled items. During this period, the focus of attention remained toward optimizing conventional or linear supply chains, less than sustainability or circular economy issues. However, the research was mostly operational and based on logistics refinement and planning, without the inclusion of environmental and circular aspects yet to be underlined in the forthcoming years.

By the mid-2010s around the period between 2015 and 2017, there was a turn toward the incorporation of circular economy concepts and a host of terminologies related to sustainability. Keywords such as remanufacturing and reverse logistics have begun to be in the central stage, with the growing importance of supply chains. Practitioners and research attention have shifted toward product reuse and material recovery, resulting in the prominence of more sophisticated mathematical tools such as stochastic programming, robust optimization, and multi-objective optimization. These techniques help address the increased complexity and uncertainty inherent in managing product returns and recycling processes [30]. This step toward circularity also called for the development of novel approaches to decision-making because old linear models were not able to capture the variability and dynamic nature of circular systems.

Sustainability, carbon emissions, and life cycle analysis have been emphasized with the highest vigor in the last five years, that is, from 2018. This trend depicts the holistic integration of environmental issues within supply chain studies, and the alignment of supply chains with sustainability objectives. During this period, the terms decision support systems, sustainable development, and circular economy became relevant [6]. This reflects a general trend from operational efficiency to environmental and social performance impacts, which affect supply chain strategies and operating models. Among other solving tools, system dynamics and simulations are increasingly being applied in modeling complex systems with multiple objectives, such as the reduction of environmental impacts and economic performance. These tools have been deployed to comprehend and optimize the complex interrelations that exist in circular supply chains involving multiple flows of resources and products.

The evolution of the field of operational research reflects the increasing complexity of managing circular supply chain networks. Although early research focused on optimizing static and predictable linear supply chains using deterministic methods, the uncertainty and variability introduced by reverse logistics, recycling, and sustainability goals have become key constraints to consider when developing new operational techniques and methods. The shift from traditional optimization techniques toward state-of-the-art methods, such as robust optimization and multi-objective decision-making, indicates that there is an ever-growing demand for flexibility and adaptability within today's supply chains. Currently, research emphasis has shifted further to decision support systems and sustainability-driven models, which also cover economic efficiency and environmental responsibility in circular supply chains.

4.5 Discussion

The integration of advanced operational research methods in circular economy frameworks has been developing rapidly in recent years as the complexity and scale of sustainable supply chains have significantly increased. Recent breakthroughs in machine learning (ML), artificial intelligence (AI), accelerated Benders decomposition, and cutting-plane algorithms continue to support traditional OR techniques, such as linear programming, dynamic programming, and heuristic methods, to further optimize decision-making and enhance the efficiency of circular processes. These have become very important in addressing the increasing demand for sustainable operational and resource optimization [30].

Traditional OR methods have played an enabling role in the development of circular economic strategies, particularly in the design of circularity, reverse logistics, and remanufacturing. For example, myriad optimization and integer programming approaches have been developed to solve resource allocation problems during the design phase by balancing material usage against long-term reuse potential. Using mathematical programming and simulation models, businesses have been able to manage the flow of returned products in reverse logistics, where there may be fluctuations in return rates and product conditions. In the same vein, stochastic programming methods have played a major role in remanufacturing, as they help address the uncertainty related to the quality and timing of returned products, whereas integer programming ensures efficiency in allocating resources during the remanufacturing process [18]. However, these traditional methods cannot stand alone with the growing complexity of circularity requirements and constraints, combined with the increasing complexity of supply chain networks. Recently developed machine learning and AI have opened new opportunities for data-driven decision-making in cases where a large amount of data is generated from IoT-enabled devices [22]. These technologies enable the realization of predictive maintenance in repair and remanufacturing while optimizing processes against real-time data. For instance, AI has been introduced to improve reverse logistics and recycling flows by learning from historical data to forecast the future demand for return items.

Several publications have also deployed the accelerated Benders decomposition method as a particularly effective method for large-scale optimization problems. It finds specific applications in network design and resource allocation within the context of a circular supply chain, where it helps to accelerate convergence in complex mixed-integer programming problems [13]. This has become important in the optimization of the location of recycling centers, reverse logistics hubs, and remanufacturing facilities in an economy shifting toward a fully circular system [23]. Accelerated Benders decomposition enhances solution efficiency and precision by partitioning large-scale problems into smaller, more manageable subproblems. This method significantly improves computational speed and solution accuracy, enabling faster convergence and optimized resource allocation. From another perspective, cutting-plane algorithms have gained momentum for solving integer programming problems related to the optimization of remanufacturing schedules and resource allocation. The procedure entailed by these algorithms iteratively refines a feasible solution region. As such, cutting-plane algorithms have immense power in dealing with multistep processes related to circularity. For instance, assume that cutting planes can be utilized to optimize the flow of recycled materials back into the supply chain in such a way that resources are reused as efficiently as possible with as little material being wasted. Traditional heuristics and metaheuristics, including, but not limited to, genetic algorithms and simulated annealing, have been widely applied to complex optimization problems in circular economies. Because the exact underlying methods of such problems are often computationally expensive or even infeasible, there are areas of application, such as recycling and reverse logistics. Metaheuristics, therefore, allow flexibility and adaptiveness to solve problems relevant to the quest for circularity, such as routing, waste sorting, and material recovery. In addition, as supply chains become increasingly exposed to uncertainty and variability, stochastic and robust optimization methods are becoming increasingly important. By nature, the processes of a circular economy are exposed to demand fluctuations, the quality of returned products, and material availability. Stochastic programming provides solutions to account for these types of uncertainties, considering the variability in the solution, whereas robust optimization ensures that the solutions remain valid under disruptions. For instance, robust optimization may be critical in remanufacturing, as the variability of returned goods quality can have a large impact on efficiency during the process [13].

Optimization, linear programming, and heuristics have proven to be valuable and traditional tools in advancing circularity, growing complexity in supply chains, and sustainability goals require even more advanced techniques. Machine learning, AI, accelerated Bender's decomposition, and cutting-plane algorithms have introduced computational efficiency and data-driven insights for solving modern-day supply chain challenges [21]. Some of these techniques have become imperative in solving far-reaching issues related to resource efficiency, reduced waste, and sustainable performance at the global economic level. Together with these advanced methods, conventional approaches to OR will help organizations create more resilient, adaptive, and efficient circular supply chains that could equip them to tackle challenges in a fast-changing world.

A major limitation of the CSCM research is the limited deployment of advanced quantitative techniques. While a few basic statistical methods, such as regression analysis and analysis of variance (ANOVA), have been used in some studies, more advanced approaches have been used, such as SEM, factor analysis, Bayesian analysis, and path analysis. These tools might help offer deeper insights into the relationships of variables in CSCM, thus providing a clearer understanding of the impact that circularity practices have on supply chain performance. The limited application of these tools constrains the development of more rigorous evidence-based research that can better inform decision-making in a circular economy. Furthermore, multi-objective programming and system dynamics have not yet gained widespread acceptance in CSCM research. These are essential for capturing the multidimensional complexity of circular supply chains. For example, the economic, environmental, and social objectives of alignment often require sophisticated optimization techniques that consider multiple, often conflicting, objectives, such as sustainability versus operational efficiency trade-offs or optimization of material flows across closed-loop supply chains.

AI and machine learning model development for the optimization of resource circularity and reverse logistics is still at an emerging stage, yet it has high potential to advance CSCM. Further research is needed on how these technologies can improve resource allocation, routing, and recovery processes. In this respect, AI-driven predictive maintenance systems, which allow prediction when repairs are needed during remanufacturing, have the potential to avoid downtime and optimize resources, which requires further research and investigation [24]. Second, research on waste sorting automation using AI has great potential for improving the material classification for recovery. Finally, establishing standardized data frameworks provides a foundation upon which AI and ML can be effectively applied to CSCM.

4.6 Conclusion

This study underlines the increasing relevance of incorporating advanced operational research techniques into circular supply chain management, considering both the growing size and increasing complexity of contemporary supply chains. Firms are moving from a linear approach to a circular one in which optimizing resource use, waste minimization, and balancing economic and social-environmental objectives become imperative. Traditional OR methods of optimization, simulation, and decision analysis have been found to be useful in circularity, particularly in reverse logistics, remanufacturing, and resource recovery. However, more advanced approaches are called for because of the challenges of sustainability, which evolve at an ever-increasing pace.

New opportunities in advancing decision-making, prediction, and control will be afforded by emerging technologies such as AI, ML, and advanced optimization algorithms, including, but not limited to, accelerated Benders decomposition and cutting-plane methods. These tools are necessary to deal with uncertainties, enhance predictive maintenance, automate waste sorting, and optimize material recovery processes. In addition, this study emphasizes how standardized data frameworks are developed to allow AI and ML to be applied to many of the processes involved in CSCM.

Although OR methods have significantly contributed toward the circular economy, the integration of more recent technologies remains a critical key toward future development. Accordingly, this study calls for further research into how these technologies and new methodologies can enhance the circular economy journey to enhance supply chain sustainability and efficiency.

References

1. Agrawal, V., Atasu, A., & Ülkü, S. Leasing, modularity, and the circular economy. (2021). Management Science, 67(11), 6782–6802.
2. Akpan, I. J. Thirty years of International Transactions in Operational Research: past, present, and future direction. (2023). International Transactions in Operational Research, 30(6), 2709–2728.
3. Barbosa-Póvoa, A. P., da Silva, C., & Carvalho, A. Opportunities and challenges in sustainable supply chain: An Operational Research perspective. (2018). European journal of operational research, 268(2), 399–431.
4. Chen, G., & Xiao, L. Selecting publication keywords for domain analysis in bibliometrics: A comparison of three methods. (2016). Journal of Informetrics, 10(1), 212–223.
5. Das, C. Synthesising and Conceptualising Circular Supply Chains: A State-of-the-Art Literature Review. (2024), The Journal of Environment & Development, 10704965241231553.
6. de Sousa Jabbour, A. B. L., Luiz, J. V. R., Luiz, O. R., Jabbour, C. J. C., Ndubisi, N. O., de Oliveira, J. H. C., & Junior, F. H. (2019). Circular economy business models and operations management. Journal of cleaner production, 235, 1525–1539.
7. Dekker, R., Bloemhof-Ruwaard, J., & Mallidis, I. (2015). A Hierarchical Decision-Making Framework for Quantitative Green Supply Chain Management: A Critical Synthesis of Academic Research Efforts. Supply Chain Management for Sustainable Food Networks, 129–157.
8. Dekker, R., Bloemhof, J., & Mallidis, I. (2012). Operational Research for green logistics–An overview of aspects, issues, contributions and challenges. European journal of operational research, 219(3), 671–679.
9. Demeester, L., Qi, M., & Van Wassenhove, L. N. (2013). Plant networks for processing recyclable materials. Manufacturing & Service operational Management, 15(4), 670–688.
10. Fleischmann, M. (2012). Quantitative models for reverse logistics (Vol. 501). Springer Science & Business Media.
11. Georgiadis, P., & Athanasiou, E. (2013). Flexible long-term capacity planning in closed-loop supply chains with remanufacturing. European Journal of Operational Research, 225(1), 44–58.
12. Gonzalez, E. D., Sarkis, J., Huisingh, D., Huatuco, L. H., Maculan, N., Montoya-Torres, J. R., & De Almeida, C. M. (2015). Making real progress toward more sustainable socie-ties using decision support models and tools: introduction to the special volume. Journal of Cleaner Production, 105, 1–13.
13. Govindan, K., Mina, H., Esmaeili, A., & Gholami-Zanjani, S. M. (2020). An integrated hybrid approach for circular supplier selection and closed loop supply chain network design under uncertainty. Journal of Cleaner Production, 242, 118317.
14. Henao, R., & Sarache, W. Sustainable performance in manufacturing operational: the cumulative approach vs. trade-offs approach. (2022). International Journal of Production Economics, 244, 108385.
15. Holmström, J., Hameri, A. P., & Ketokivi, M. (2006, August). Operational MANAGEMENT AS A PROBLEM-SOLVING DISCIPLINE. In Academy of Management Proceedings (Vol. 2006, No. 1, pp. B1-B6). Briarcliff Manor, NY 10510: Academy of Management.
16. Lahane, S., Kant, R., Shankar, R., & Patil, S. K. (2024). Circular supply chain implementation performance measurement framework: A comparative case analysis. Production Planning & Control, 35(11), 1332–1351.
17. Lee, Y., Hu, J., & Lim, M. K. (2024). Revisiting circular economy indicators: A circular supply chain perspective. Journal of Purchasing and Supply Management, 100941.
18. Liao, H., Zhang, Q., Shen, N., & Li, L. (2020). Stochastic analysis of quality uncertainty and optimal acquisition strategies for engine remanufacturing. Journal of Cleaner Production, 261, 121088.

19. Luthra, S., Sharma, M., Kumar, A., Joshi, S., Collins, E., & Mangla, S. (2022). Overcoming barriers to cross-sector collaboration in circular supply chain management: a multi-method approach. Transportation Research Part E: Logistics and Transportation Review, 157, 102582.
20. Nahmias, S. (1981). Approximation techniques for several stochastic inventory models. Computers & operational Research, 8(3), 141–158.
21. Nasr, A. K., Tavana, M., Alavi, B., & Mina, H. (2021). A novel fuzzy multi-objective circular supplier selection and order allocation model for sustainable closed-loop supply chains. Journal of Cleaner production, 287, 124994.
22. Noman, A. A., Akter, U. H., Pranto, T. H., & Haque, A. K. M. (2022). Machine learning and artificial intelligence in circular economy: a bibliometric analysis and systematic literature review. Annals of Emerging Technologies in Computing (AETiC), 6(2), 13–40.
23. Roghanian, E., & Pazhoheshfar, P. (2014). An optimization model for reverse logistics network under stochastic environment by using genetic algorithm. Journal of Manufacturing Systems, 33(3), 348–356.
24. Sadeghi, R. K., Abadi, M. Q. H., Haapala, K. R., & Huscroft, J. R. (2024). A hybrid Machine learning solution for redesigning sustainable circular energy supply chains. Computers & Industrial Engineering, 110541.
25. Sinuany-Stern, Z. (2023). Foundations of operational research: From linear programming to data envelopment analysis. European Journal of Operational Research, 306(3), 1069–1080.
26. Thies, C., Kieckhäfer, K., Spengler, T. S., & Sodhi, M. S. (2019). Operational Research for sustainability assessment of products: A review. European Journal of Operational Research, 274(1), 1–21.
27. Van Loon, P., & Van Wassenhove, L. N. (2020). Transition to the circular economy: the story of four case companies. International Journal of Production Research, 58(11), 3415–3422.
28. Vuori, N., Burkhard, B., Laamanen, T., & Bingham, C. (2024). Heuristics in Organizations: Toward an Integrative Process Model. Academy of Management Annals, 18(2), 670–711.
29. Zhang, A., Wang, J. X., Farooque, M., Wang, Y., & Choi, T. M. (2021). Multi-dimensional circular supply chain management: A comparative review of the state-of-the-art practices and research. Transportation Research Part E: Logistics and Transportation Review, 155, 102509.
30. Zils, M., Howard, M., & Hopkinson, P. (2023). Circular economy implementation in operations & supply chain management: Building a pathway to business transformation. Production Planning & Control, 1–20.

Chapter 5
Theory and Research Concerning the Circular Economy Model and Future Trend

Harun Resit Yazgan, Derya Eren Akyol, and Gulseli Altintaş

Abstract This chapter aims to draw attention to the Circular Economy approach, address a future trend, and create a compilation of important Circular Economy studies published between 1994 and 2023. The failure to solve problems such as climate change, loss of biodiversity, overpopulation, and resource scarcity by the current Linear Economy model has shown that sustainable economy models can be a successful solution. On the other hand, there are also emerging technology trends that have the enormous potential to impact Circular Economy. The first contribution of this study is the investigation of the impact of future trends on Circular Economy. Three trends are selected and suggested to be examined: Artificial Intelligence, Industry 4.0, and Operational Research approaches. We collected studies (published between 2007 and 2023) examining Circular Economy from different perspectives in a single source and also highlight the usability areas of selected approaches that integrate the Circular Economy into production. Another contribution to the literature is to propose the use of Artificial Intelligence regarding Circular Economy's 9R framework. The study guides the implementation of the Circular Economy concept and provides a significant contribution to the Circular Economy literature from different perspectives.

Keywords Sustainability · Circular economy · Product life cycle · Artificial intelligence · Industry 4.0 · Operational research

H. R. Yazgan · G. Altintaş (✉)
Engineering Faculty, Industrial Engineering, Sakarya University, Serdivan, Turkey
e-mail: yazgan@sakarya.edu.tr; gulseliisler@sakarya.edu.tr

D. E. Akyol
Engineering Faculty, Industrial Engineering, Dokuz Eylul University, Konak, Turkey
e-mail: derya.eren@deu.edu.tr

B. Jarboui et al. (eds.), *Circular Economy Supply Chains*,
Computational Intelligence Methods and Applications,
https://doi.org/10.1007/978-981-96-7905-8_5

5.1 Introduction

Our nature consists of processes that involve an endless cycle, such as a waste-free metabolism. Since an input of one process is the output of another process in nature, no waste is generated. The Circular Economy (CE) model, which is seen as an innovative alternative to the frequently used Linear Economy (LE) model, is a new and advanced system that optimizes natural resources used and keeps them in product life cycles as long as possible. Also, the CE model transforms products that have completed their life cycles into sources of other product and service cycles [91]. There is no clear evidence of the source or originator of the CE concept. Most inspirations in the literature are based on the Industrial Ecology Model based on the closed-loop concept originating from Japan and Germany [53, 101]. However, the following names can be seen among those who contributed to the emergence of this concept: [49, 51, 79, 97]. While Jawahir and Brandley [35] and Homrich et al. [29] claim that the term CE was first used by British economists Pearce and Turner in 1990, but Pearce and Turner [87] claim that this term was first used by Stahel and Reday-Mulvey [80] in the Western literature in the 1980s to denote a system in which environment and economy have closed interactions [21]. Some sources state that the CE concept emerged with a report from Switzerland presented to the European Commission by Geneviève Reday-Mulvey and Walter R. Stahel [81], explaining the idea of using human power instead of energy. We present a literature review on the origin, logic, structure, and principles of CE and give detailed explanations. Differences between CE and LE are stated, and key studies (published between 1994 and 2023) on sustainability and CE worldwide are included. In addition, principles, processes, potential benefits, and potential barriers of the CE are listed. A literature review on the relationship between CE and the product life cycle is included. In literature, business model strategies of CE are explained as a preliminary preparation for adapting the CE to the product life cycle. On the other hand, there are also emerging technology trends that have the enormous potential to impact CE. To investigate the impact of future trends on CE, three trends are selected and suggested to be examined: AI, Industry 4.0, and OR approaches. AI is an emerging technology with enormous potential to impact CE [95]. AI has the potential to enable and enhance the CE by providing new ways to optimize material and energy flows, reduce waste, and increase resource efficiency. In this context, the studies (published between 2007 and 2023) from the literature were examined, and the usage areas of AI in applying principles for transitioning to CE in product life cycle processes were proposed. Furthermore, this study shows how AI can be used in terms of the 9Rs (Refuse, Rethink, Reduce, Reuse, Repair, Refurbish, Remanufacture, Repurpose, Recycle, and Recover) framework of CE. The transformative tools of Industry 4.0 technologies align with CE principles that focus on reducing waste and maximizing resource efficiency. Industry 4.0 technologies such as smart manufacturing, data analytics, and the Internet of Things (IoT) enable companies to easily monitor and optimize their resource usage, product life cycles, and supply chains, in line with the CE concept. The interactive and combined use of CE and Industry 4.0 can enable economic growth to progress in harmony with environmental responsibility. In this context, important studies from the literature dealing with Industry 4.0 within the scope of CE are included and examined.

Operational Research (OR) models have gained significant popularity in diverse fields due to their ability to provide data-driven insights and optimize complex decision-making processes. OR approaches are in a symbiotic relationship with the CE model, and OR acts as a facilitator for the effective implementation of circular applications. Using OR, companies can manage the complexities of CE with precision and achieve a combination of environmental sustainability and economic efficiency. In this context, some important studies from the literature dealing with OR within the scope of CE are included and examined.

5.2 Material and Method

5.2.1 Circular Economy Model Concept and Comparison with Linear Economy

LE, i.e., "take-make-dispose" model that has shaped the global economy since the industrial revolution, is a one-way system in which natural resources turn into waste through production. Since this system is based on the principle that natural resources in our world are easily accessible, sufficient, and available and become cheap and trouble-free waste at the end of their lifespan, we have faced a serious increase in the risk of world resource scarcity in many sectors (Steffen et al., 2015). Other factors that make this model unsustainable are the high demand for resources and the increase in population, inequality, demographic change, and consumption. Globalization has led to increased complexity in supply chains and production processes, along with greater price fluctuations and heightened competition in both national and local markets. In addition, consequences such as the COVID-19 pandemic, rapid depletion of natural resources, and climate change push us to change the ecological boundaries of our world. It is seen that the economic model (LE) being used worldwide cannot prevent these crises and cannot meet the needs of humanity.

Today, humankind consumes 1.5 times the regenerative capacity of the Earth's natural ecosystems. Globally, our consumption exceeds nature's capacity by 50%. If our consumption patterns do not change, it is projected that by 2050, with the increasing population, consumption will triple (TEMA Foundation, 2015 [88]). As a result of this increase, it is predicted that global resource use will increase by more than three times by 2050. We cannot adequately manage waste generated due to the resources we use: Globally, 1/3 of plastic waste is not collected or managed, and greenhouse gas emissions have increased over the years (United Nations Environment Programme, 2021 [89]).

CE logic is different from LE. This concept transforms linear production into a competency model: It follows a path of maximum reuse, recycling of nonreusable goods, repair of damaged, etc., and reproduction of parts and products that are beyond repair [43].

The CE model is a model and tool with a holistic structure and a production system that enables the reuse of raw materials and semi-finished and finished products [31]. At the same time, in a CE, wastes are recycled into the system, resources and energy are used very efficiently, and waste and environmental harm are minimized. The CE approach aims to maximize the added value of services and products in the value chain and ensure that resources are in the cycle and the economy for much longer [53].

The European Union (EU) wants to reduce raw material production requirements, to obtain economic and environmental benefits, thanks to the transition to a new economic concept that adopts renewable energy and efficient use of resources. For this purpose, the CE package was adopted by the European Commission on December 2, 2015, which sets out legislative proposals on waste and a comprehensive action plan for a circular and sustainable economy. In line with this plan, some countries that are members of the EU have started to prepare and accept regional and national roadmaps for the transition from LE to CE [72]. According to the Directive 2000/53/EC presented by the European Parliament and of the Council [18], starting from 2015, the target is to attain a percentage of 95% in terms of the total mass of reused and recovered materials, and at least 85% of this percentage is to be obtained from reuse and recycling [57].

Furthermore, the CE, adopted by the "European Green Deal" ([20], is one of the priority policy areas in the new strategy. The European Green Deal, approved in 2020, is a set of policy initiatives by the European Commission with the overarching aim of making the European Union (EU) climate-neutral in 2050. With this document, 2050 is targeted for the EU to offer new and better jobs, restore climate neutrality, improve people's well-being, and increase growth. At the same time, the EU started the work needed to use resources more efficiently, stop biodiversity loss and climate change, and reduce environmental pollution by transitioning to a circular, i.e., clean economy. In the last decade, the United States, China, and South Korea have introduced research programs that promote circular economies in product and service markets by increasing reuse.

The 2030 Agenda for Sustainable Development was adopted by the United Nations General Assembly on September 25, 2015 [90], thus taking a step toward implementing the CE model. In line with these studies, the EU's objectives for production are to develop some incentives that support cyclical product designs and to create efficient and innovative production-product life cycle processes. Strategies for recycling include promoting recycled raw materials, nutrients, and water, increasing knowledge about material flows in the cycle, and ensuring safe chemical management.

CE renews and repairs products and services through design. The overall goal of the CE is to maximize the value of a product at every point in its life cycle. This economy aims to optimize returns of resources, develop and protect natural capital, and minimize risks in a system with flow chain management [74]. It delivers effective results in businesses of all sizes [91].

5.2.2 *Circular Economy Structure*

The CE is achieved by designing all industrial processes so that resources, services, and raw materials do not leave the product life cycle and are used continuously. In this direction, while wastes are minimized, residues that cannot be removed from the cycle are also recovered [91]. In a CE process, raw materials, components, and products are used in biological or technical cycles. In contrast, environmental impacts of services and products are minimized within the product life cycle/value chain. The diagram described by the Ellen MacArthur Foundation [13] includes two cycles of the CE (it is shown in Fig. 5.1):

1. The biological cycle where residues and wastes are recycled to nature after use
2. The technical cycle in which raw materials, components, and product designs are made to minimize waste

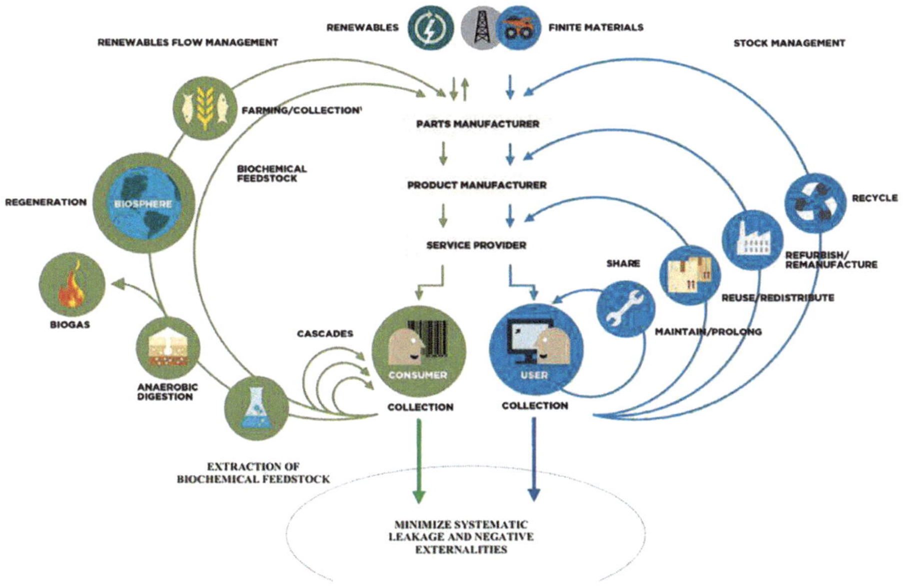

Fig. 5.1 CE diagram [15]

5.2.3 Benefits and Barriers of Applying Circular Economy to Product Life Cycle

5.2.3.1 Potential Benefits of CE

The CE is an economic concept that provides new strategies and business models to reuse resources with maximum efficiency and keep them in cycles within the confines of our planet. In this context, implementing the CE brings many benefits to businesses, countries, and our planet (in terms of sustainable use of its resources).

CE, a new concept, is seen to provide many economic, social, and environmental benefits. J. Korhonen et al. [44] group the dimensions in which sustainable development and implementation of a successful CE benefit under three headings: economic, social, and environmental.

Making a radical and fundamental change and renewal in business processes, models, products, and factory designs is a huge step for a company. However, it is stated that this change will lead the company to the result of commercial success, and it will be an incentive factor for other companies [17]. There are observations and comments in the literature about the benefits of the CE. The benefits of applying CE are:

- Globally 4.5 trillion dollars growth opportunity (Lacy et al., 2020: 3) [47]
- Increase in global transparency and collaboration
- Extending product life cycles
- Ease of disassembly and repair
- Establishment of new and innovative business models
- Creating new values for materials that have already been paid for (Wijkman and Skånberg, 2015) [94]
- Minimization of costs (Ünlütürk et al., 2020) [91]
- Efficiency opportunities in supply chains and industry in general (Wijkman and Skånberg, 2015) [94]
- Widespread use of reusable, compostable, and recyclable parts, resources, and products
- Saving by reducing dependence on energy and primary raw materials
- Ensuring the sustainability of the export of products and services (Ünlütürk et al., 2020) [91]
- Opportunity to increase modularity of products
- Increase in functionality of products
- Increase in upgradability of products and services
- Growth despite using fewer resources (Ünlütürk et al., 2020) [91]
- Increase in security of supply
- Minimizing price fluctuations in raw materials (Ünlütürk et al., 2020) [91]
- Controlling the ever-increasing costs of purchasing, production, transportation, etc.
- Contributing to the concept of combating climate change as a result of reducing greenhouse gas emissions, recycling water and food, and waste management

- New employment and job opportunities (Esposito et al., 2015) [17]
- Reducing the environmental impacts of increasingly intense resource use and waste uncontrolled, resulting from the actions of humanity so far
- Increase in annual economic incomes nationally and locally
- Increase in technological developments
- Strengthening design and innovation as businesses aim to transition to a CE (Bocken et al., 2016) [5]
- Increase in the competitiveness, flexibility, and endurance of businesses (Ünlütürk et al., 2020) [91]
- Chance of innovation and R&D to create new opportunities (Ellen MacArthur Foundation, 2019) [14]
- Greatly reducing greenhouse gas emissions and primary resource consumption in the short, medium, and long terms
- Creating new productivity and labor opportunities in supply chains and industry (Wijkman and Skånberg, 2015) [94]
- Providing economic value by reducing the ecological footprint and raw material costs and minimizing price variability (Buruzs and Torma, 2017) [6]
- Opportunity to improve resource recovery systems and asset sharing systems through new technologies (Hopkinson et al., 2018) [30]
- Increasing the benefits of reverse logistics application through new technologies suitable for CE (Hopkinson et al., 2018) [30]

5.2.3.2 Potential Barriers of CE

The CE concept ensures that the life of products is extended by developing improvement, recycling, and repair processes instead of making waste after use. In fact, barriers to the implementation of a CE are similar to the barriers to the integration of sustainability strategies [67]. While the transition to the CE requires massive and radical changes in the production and consumption processes, it also has barriers at micro and macro scales.

Jesus and Mendonça [34] classified the barriers to CE as soft barriers (institutional and social) and hard barriers (technical and economic). Kirchherr et al. [43] characterized the barriers to CE as legal, technological, market, and cultural, while Hart et al. [28] characterized them as sectoral, legal, financial, and cultural. Moreover, Grafström and Aasma [24] classified the barriers to the CE as technological, economic and market-based, legal and institutional, cultural, and social.

Although these barriers are written as separate items in the literature, it is stated that all these are related to one another. Some evaluations of removing a barrier will be a factor (catalyst-like) in removing barriers under other headings [24].

The list of barriers to the use of the CE is:

- The necessity of breaking down existing production and consumption patterns (Haas et al., 2015) [26]

- Structural designs of almost all existing products on the current market that are not suitable for recycling or reuse
- Lack of clear plans for the recovery of products, although the recovery flow from the market for reuse or recycling of products needs to be improved
- Insufficiency of existing industrial infrastructure for recycling and reuse
- Difficulty in disassembling parts of waste products that are in good condition
- Risk of damaging dismantled parts of the product that is in good condition when disassembled from the product
- The fact that maintenance, repair, and refurbishment of products manufactured according to LE, which is widely used in the market, are currently costlier than buying a new product
- The necessity to establish a return flow, reuse and redistribution network for disassembled products, their parts, and recycled materials
- The perception that the quality of remanufactured or recycled materials to be used as inputs to new systems will be low (Andrews, 2015) [3]
- The necessity of planning and studies to encourage investments
- The necessity of developing management strategy and skills in realizing radical innovations and implementing disruptive changes
- The possibility of the CE model to change cultural structures and break the resistance in the structure
- The necessity of the advanced technological developments that the model will bring to minimize costs (Kirchherr et al., 2018) [43]
- The obligation to establish the necessary information systems infrastructure to monitor and control progress, steps, and results
- The fact that the CE logic based on permanence does not comply with the fashion concept that determines today's trends
- The fact that existing and potential customers are not open to rapid and radical changes
- Lack of sufficient awareness among current and potential consumers
- The necessity of marketing/encouragement studies to eliminate lack of enthusiasm in current and potential consumers and to change habitual behaviors
- The necessity of increasing cooperation with suppliers
- The fact that there is no recognized and independent organization yet, despite the long-term necessity of applications such as labeling and certification
- Lack of standardization of certification practices in the international market yet (Preston, 2012) [65]
- Although a transition to renewable energy is necessary to reduce fossil fuel consumption, intensive use of fossil fuel consumption for energy production

5.2.4 Circular Economy Business Models

Changeover to the CE model requires many innovative and fundamental changes and improvements. In this part of the study, business model concepts are presented

Table 5.1 CE business models

Business model strategy	Scope
CE Business Model Strategies Aimed at Slowing the Cycles	Business models that extend product or service life/encourage reuse; support the CE through repair, remanufacturing, upgrades, and retrofits
CE Business Model Strategies Aimed at Closing the Cycles	Business models that enable recycling of products and services; transform old/refused products and parts into new resources

by Bocken et al. [5] that fit the approaches to slowing down resource and closing cycles. As can be seen in Table 5.1, CE business models are broadly divided into two groups.

5.2.4.1 Circular Economy Business Model Strategies to Slow Cycles

CE business model strategies to slow cycles are business models that extend product or service life and encourage reuse. These business models support the CE through repairs, remanufacturing, upgrades, and improvements. CE business models for slowing cycles are listed below [5].

Access/performance (sharing) business model: It is a business model that adopts a method of meeting the usage needs of consumers without owning physical products.

Business model aiming to increase product value: Uncovering and using the value of waste materials is a business model that adopts a method of product residues generally in the form of production-consumer production.

Long life (classic) business model: It is a business model that adopts a method of designing products to allow repair and provide durability support, thereby increasing the product's lifetime.

The business model aimed at supporting competence: It is a business model that adopts a method of developing and implementing solutions (such as service, durability, warranty and repair, and upgradeability) that will reduce the consumption of end consumers.

5.2.4.2 Circular Economy Business Model Strategies to Close Cycles

CE business model strategies to closing the cycles are business models that provide recycling of products and services. In business models that aim to close product and service life cycles, closing the cycle between supply/production and end of use is the basic element [71]. These business models enable the CE by transforming old/refused products and parts into new resources. CE business models for closing cycles are listed below [5].

The business model aimed at extending resource value: It is a business model that includes recycling products, collecting resources and materials that are seen as waste and thought to go to waste, etc., adding value to them by putting them in a new form.

Industrial symbiosis business model: It is a business model for geographically close processes that involve using residues of one as a resource for the other and introducing them into the product life cycle.

5.2.5 Circular Economy and Future Trends

This chapter examines future trends within the scope of the CE concept. In this study, three future trends, which are among the most popular trends today, were examined: AI, Industry 4.0, and OR models.

AI has grown in popularity in recent years. The popularity of AI in the manufacturing industry has grown as industries realize the potential of AI to increase productivity and innovation. In manufacturing, AI systems can streamline production lines, reduce errors, and improve product quality. Predictive maintenance models can predict equipment failures, minimizing downtime, and AI-driven analytics can optimize supply chain logistics. In addition, AI-powered quality control processes ensure consistency and reliability in output. As the manufacturing industry continues to embrace AI, this growing reliance on AI technologies in the manufacturing industry has a transformative effect, enabling companies to stay competitive, reduce costs, and bring superior products to market.

On the other hand, the fourth industrial revolution covers the use of digital technologies in human life, including AI, the Internet of Things (IoT), big data analytics, and automation. The popularity of Industry 4.0 in the manufacturing sector has reached high levels. Characterized by the integration of digital technologies, automation, and data-driven decision-making, Industry 4.0 has revolutionized manufacturing and will continue to do so. For greater efficiency, flexibility, and rapid adaptation to market demands, industries are using Industry 4.0, from the Internet of Things (IoT) to big data analytics and augmented reality applications.

OR models have gained significant popularity in diverse fields due to their ability to provide data-driven sights and optimize complex decision-making processes. These models, based on mathematical optimization techniques, help manufacturers make data-driven decisions regarding production schedules, inventory management, and distribution networks. By harnessing the power of these models, manufacturers can minimize waste, reduce lead times, and improve resource allocation, resulting in cost savings and enhanced overall efficiency. As businesses increasingly recognize the value of OR approaches in solving complex problems and enhancing decision quality, their popularity continues to grow. Table 5.2 shows key studies published between 2007 and 2023 that have considered these three future trends in the context of CE.

5.2.5.1 Relationship Between Circular Economy and Artificial Intelligence

The relationship between CE and AI is an important and emerging area of research and development. The CE is an economic model that aims to minimize waste and maximize resource efficiency while creating a closed-loop system where materials are reused, repurposed, and recycled. AI, on the other hand, deals with developing intelligent machines that can perform tasks that typically require human intelligence. AI is an emerging technology with enormous potential to impact CE [97]. The combination of AI and the CE model presents exciting opportunities for sustainable innovation and growth. According to our limited knowledge, there is no study regarding the CE processes where AI can contribute to the CE model and shape future trends. This section explores the use of AI in integrating CE into product life cycle processes. There are studies using AI within the scope of CE and Sustainability in different sectors.

Weichhart et al. [93] addressed the need to create a true sensing and smart enterprise that aligns with sustainability requirements. It emphasizes the necessity for methods to capture the reality of enterprises and establish an interoperable digital enterprise model. The study outlines various challenges in enterprise modeling that should be addressed in future research. Additionally, it introduces the concept of the Sensing, Smart, and Sustainable Enterprise System.

Abou Baker et al. [1] discussed the use of AI in efficiently sorting and recycling e-waste, emphasizing the intersection of AI and CE. This study highlights the importance of automated e-waste recycling due to the increasing volume of e-waste. They specifically focus on smartphone classification, using transfer learning with the AlexNet model. They fine-tune the model, optimize learning parameters, and augment the dataset to achieve nearly 98% accuracy.

Kumar et al. [46] propose an AI-based solution to automatically sort COVID-related medical waste from other types and make data-driven recycling decisions. The solution relies on image-texture-dependent features and a support vector machine (SVM) classifier, achieving high accuracy (96.5%) in waste classification, particularly for circular manufacturing and addressing the issue of mixed COVID-related medical waste.

Wilts et al. [96] discussed a research project called "ZRR for municipal waste" that aimed to automate municipal waste sorting plants by using robots with AI (ZRR) to supplement or replace manual sorting. The project's objectives included increasing recycling rates, improving the purity of recovered materials, recovering additional materials from rejected flows, and enhancing the working conditions of employees who could focus on robot maintenance. The paper presents the main findings from the project, which represents the first attempt to apply AI for separating bulky municipal solid waste (MSW) in a full-scale waste treatment plant.

Andono et al. [2] suggested that addressing supply chain and customer relationship management challenges through an integrated system can help balance supply and demand, particularly for micro, small, and medium enterprises (MSMEs) in the agricultural value chain in Indonesia. This integration can potentially reduce the need for importing agricultural commodities, which is currently driven by significant demand but negatively impacts local farmers' yields and onion prices. The study focuses on modeling the optimization of the entire onion value chain in Indonesia, including aspects like land use, harvest cycles, supply and demand, storage, delivery, and pricing. Ultimately, this optimization has the potential to enhance food security in the country.

Pregowska et al. [64] addressed the challenge of managing battery waste, particularly recycling, to preserve natural resources and ecological balance. It emphasizes the role of AI in cost reduction, automation, and productivity enhancement. The paper reviewed the current state of AI-based battery waste management research, focusing on trends, algorithm accuracy, and data availability within the context of sustainable development.

The objective of Neto et al. (2023)'s [11] study was to apply simulation and computational intelligence techniques using AI and genetic algorithm for economic and environmental optimization of the reverse network (manufacturers, waste managers, and recyclers in São Paulo, Brazil) of waste electrical and electronic equipment (WEEE) to promote CE. The study focused on two main objectives: economic optimization by reducing various costs and environmental evaluation by assessing impacts on different environmental components. The optimized WEEE reverse chain in São Paulo successfully reduced the number of collections and maximized storage space utilization, indicating potential economic and environmental benefits in support of CE practices.

Onyeaka et al. [56] suggested that AI can play a pivotal role in addressing this problem and advancing the CE. The article discusses the current status of food waste and the CE and outlines how AI can be applied to monitor and optimize food production and supply chains, redistribute surplus food to those in need, and support CE initiatives. Ultimately, these AI applications aim to enhance resource efficiency, reduce environmental impact, and create a more sustainable and equitable food system. These important studies are summarized in Table 5.2.

The following strategies and principles for the transition to CE in product life cycle processes in the manufacturing sector are considered: (9R) [73, 91].

Refuse Converting the existing LE concept product into new and innovative products and services or pullback the existing product from circulation. AI can help identify products that no longer fit for purpose or are harmful to the environment and need to be withdrawn from the market [10]. By analyzing data on product safety, environmental impact, and customer feedback, AI can identify products that need to be recalled or withdrawn, enabling more sustainable and responsible business practices.

Rethink Working toward frequently using the new product developed in line with the Innovative and CE, launching it, developing reuse, and sharing models. AI can personalize products and services, making them more attractive and relevant to customers. By analyzing data on individual customer behavior, using AI, companies can design products that meet specific customer needs and preferences and develop customized reuse and sharing models that encourage greater product use and engagement. By analyzing data on consumer behavior using AI, companies can identify the most effective marketing channels and messaging that resonate with target audiences, thereby increasing awareness and adoption of circular products and services [36].

Reduce Increasing efficiency by reducing the consumption of resources and energy used during production or use of the product. With AI, by analyzing data in the production process, including machine performance, energy use, and material waste, improvement areas can be identified, and production processes can be optimized to minimize waste and reduce energy consumption [54]. AI can analyze data on energy usage, including peak demand times and patterns; thus, companies can optimize energy use and reduce costs while minimizing environmental impact. It can be used to optimize the supply chain, reduce resource consumption, and predict when equipment and machinery will need maintenance, reducing downtime and optimizing resource usage.

Reuse Reuse of a product that has not yet become a waste and is designed to be reused. Developing new business models for reuse with AI may be easier than it is now. By analyzing data on customer behavior and market trends using AI, companies can identify new opportunities for reuse, such as rental, leasing, or subscription-based models, which can help extend the life of products and reduce waste. Also, AI can help optimize the design of products for reuse and track the location and condition of assets designed for reuse by using sensors and tracking technology [68]. And also it can be used to optimize the logistics of reuse. Furthermore, integration of AI with CE can promote sustainability and waste reduction by supporting circular business models such as renting or sharing products instead of traditional selling in practice.

Repair To ensure that the product that is broken or unable to function is maintained and repaired. By analyzing product performance and usage patterns, AI can predict when a product will likely break down or require maintenance (predictive maintenance) [70]. By analyzing data from sensors and other sources using AI, product problems can be easily diagnosed remotely, and the correct parts needed to repair a product can be identified. It can be used to optimize repair scheduling by analyzing data on repair times and availability.

Refurbish Renewing/updating the product to a determined high-quality level. AI can be used to analyze products to determine the potential for refurbishment by analyzing product specifications, usage history, and performance data. AI can optimize the refurbishment process by analyzing refurbishment times, costs, and quality data. AI can help to ensure the quality of refurbished products and to personalize refurbished products for customers [22].

Remanufacture Maintaining usable parts of waste and scrap products and reusing them in a new and original product. Using AI, waste and scrap products can be analyzed to identify usable parts and components that can be disassembled and reused using machine learning algorithms to identify these parts. Once the usable parts were disassembled, by analyzing data on product specifications and performance, AI can test and sort them based on their quality and suitability for reuse [8]. It can match the usable parts with new product designs that require similar or identical components. By analyzing product specifications and performance data, AI can verify that the remanufactured products meet the required quality standards, reducing the risk of defects and ensuring customer satisfaction.

Repurpose Reuse waste and scrap products or parts in some new products. AI can help support the Repurpose model in a CE by analyzing waste and scrap products, identifying usable materials, sorting and preparing materials for reuse, designing new products that use repurposed materials, and optimizing the manufacturing process for the repurposed materials. By repurposing waste and scrapped products or parts in new products, companies can reduce waste, conserve resources, and minimize the environmental impact of their operations [25].

Recycle Ensuring that waste, scrapped products, and parts are converted into new items, parts, and products in new products or through reprocessing. AI can identify the materials in the waste, scrapped products, and parts by using machine learning algorithms to identify these materials [79]. After identifying the materials, AI can sort and separate them based on their properties and characteristics. In addition, with AI, data on production parameters and performance can be analyzed, which can be used to optimize the recycling process. AI can create new and original products that use recycled materials efficiently and effectively.

In conclusion, the CE model and AI represent a promising approach to creating a more sustainable and equitable future. The CE model and AI have the potential to create a more sustainable and prosperous future. As more businesses and governments adopt CE principles and AI technologies continue to advance, we can look forward to a more sustainable and equitable future.

5.2.5.2 Relationship Between Circular Economy and Industry 4.0

The relationship between CE and Industry 4.0 represents a pivotal intersection at the forefront of sustainable business practices. Industry 4.0 provides the means to carry out the vision of a CE, which prioritize reducing waste and optimizing resource use. This symbiotic relationship has the potential to revolutionize the production process. Smart manufacturing, data analytics, and the Internet of Things (IoT) technology allow organizations to track and optimize their resource usage, product life cycles, and supply chains in an unprecedented precision. CE and Industry 4.0 can work together to create a sustainable future in which economic growth is balanced with environmental protection. There are studies using Industry 4.0 within the scope of CE and sustainability in different sectors.

Visich et al. [92] focused on the utilization of Radio Frequency Identification (RFID) in closed-loop supply chains. They discussed how RFID technology can benefit different participants in such supply chains, enabling better decision-making during product returns and enhancing value recovery options. The study also offers guidance on implementing RFID systems within closed-loop supply chain operations.

Baccarelli et al. [4] suggested that combining Fog Computing (FC) and Internet of Everything (IoE) could create new opportunities in areas like Internet of Energy, Smart Cities, Industry 4.0, and Big Data Streaming. They introduced a novel technological paradigm called the Fog of Everything (FoE), which integrates FC and IoE to explore these possibilities.

Kim et al. [41] proposed a system framework for managing closed-loop supply chains using the Internet of Things (IoT), with a focus on radio frequency identification (RFID) technology, specifically the Gen2 RFID protocol. The study also designs an information system that includes RFID tag encoding to support the proposed framework. The primary goal is to create a system that enhances the profitability and environmental sustainability of closed-loop supply chain operations by gathering real-time information throughout the supply chain and product life cycle.

Moreno et al. [52] suggested transitioning from a linear product life cycle to a circular one, emphasizing increased energy efficiency and reduced resource usage. The study explores how Re-distributed Manufacturing (RdM) and a Product-Service System (PSS) approach, combined with digital manufacturing technologies, can facilitate this shift. A case study in the shoe manufacturing industry, called ShoeLab, is used to demonstrate this new approach and highlight the potential for circular production through RdM and 3D printing. The study also suggested that employing IDEF0 modeling can optimize RdM for reduced material, energy, and waste in manufacturing and transportation processes.

Jabbour et al. [33] advocated for integrating the CE with large-scale data (LD or big data). They presented an integrative framework that offers valuable socio-technical insights, benefiting academics, practitioners, managers, and policy decision-makers interested in advancing sustainability and data-driven approaches.

Mastos et al. [50] aimed to provide evidence of the impact of an IoT solution on the sustainable supply chain management (SSCM) performance. They presented a case study involving a scrap metal producer in the lift industry and a waste management company. The study illustrates how the implementation of an Industry 4.0 solution can enhance sustainability at both the firm and supply chain levels.

Chauhan et al. [7] identified and analyzed seven criteria related to a smart healthcare waste disposal system with a focus on CE principles to recover value from disposables. The criteria are assessed using the Decision-Making Trial and Evaluation Laboratory (DEMATEL) method. Overall, the study presented a causal relationship model that connects drivers of Industry 4.0 and CE principles to develop an innovative smart healthcare waste disposal system with CE benefits.

Tavana et al. [83] focused on Circular Supply Chain (CSC) networks, which promote sustainability and socially responsible practices through recycling, harvesting, and refurbishing. They introduced a Lagrangian relaxation (LR) algorithm to address location-inventory-routing (LIR) problems within multi-period and multiproduct CSC networks featuring heterogeneous vehicles. Additionally, the study proposes an Artificial Internet of Things (AIoT)-enabled sustainable CSC model, aiming to enhance network performance and establish a secure and traceable environment. These important studies are summarized in Table 5.2.

Essentially, CE and Industry 4.0 complement each other by leveraging technology and data to minimize waste, close resource loops, and promote sustainable practices. As businesses increasingly adopt both approaches, they not only reduce their environmental footprint but also find new avenues for cost savings, innovation, and competitiveness in an increasingly resource-constrained world. This symbiotic relationship is a driving force behind the transition toward more sustainable and circular industrial systems.

5.2.5.3 Relationship Between Circular Economy and Operational Research Models

While the CE concept is undoubtedly environment friendly, it also presents complex challenges for businesses seeking to implement circular practices efficiently. This is where Operations Research (OR) models come into play, acting as indispensable tools in optimizing and managing the intricate processes associated with the CE.

Operational Research, often referred to as OR, is a discipline that employs mathematical and analytical methods to solve complex decision-making problems. It encompasses various techniques such as linear programming, integer programming, simulation, and optimization, all of which can be applied to address the multifaceted issues posed by CE practices. CE initiatives often involve intricate supply chains, product life cycle assessments, and sustainable product design, all of which can be optimized through OR techniques. Furthermore, OR can aid in designing closed-loop systems that prioritize the continuous use and reutilization of materials. Studies that carry out OR approaches on sustainability and CE in the manufacturing sector are included.

Many studies developed a single-objective linear model. While some of these studies generally covered the production/manufacturing sector [16, 39, 63, 98], there are also some optimization studies in the digital industry [58], electrical and electronics industry [100], food and beverage industry [84]. Pishvaee et al. [63] proposed a mixed integer linear programming aimed at cost minimization for a robust optimization approach to closed-loop supply chain network design problems under uncertainty. Ene and Öztürk [16] proposed an open-loop reverse supply chain network model to maximize the design's total profit. Mixed integer linear programming with a deterministic approach was developed for the closed-loop supply chain network design problem by Kaya and Alumur Alev [39]. Wu et al. [98] compared interactive fuzzy programming approaches for closed-loop supply chain network design. Ozceylan [58] developed a mixed integer linear programming model for simultaneously optimizing closed- and open-loop supply chain networks with common components. Torgul et al. [84] suggested a network design and mixed integer nonlinear programming model in the agri-food supply chain. Yildizbasi and Arioz [100] developed a novel hybrid network optimization model in the electrical and electronics industry with a CE perspective.

Many studies proposed and applied a nonlinear programming approach in sectors such as the electrical and electronics industry [60], the network marketing industry [27], and the tire industry [60, 99] developed a mixed integer nonlinear programming for modeling and optimizing the integrated problem of closed-loop supply chain network design and disassembly line balancing. A mixed integer nonlinear programming for optimization model for traceable closed-loop supply chain networks was developed by Hajipour et al. [27], Yazgan et al. [99] systematically designed a process for collecting and transporting end-of-life tires (ELTs) to recycling facilities using the concept of Logistic 4.0.

Many studies developed a multi-objective linear model. While some of these studies generally covered the production/manufacturing sector [61], there are also some optimization studies in the public and private sectors [23], energy industry [9], agriculture industry [32, 55], textile industry [19], mining industry [78], electronics industry [48, 82], food industry [69], and battery industry [86]. Ozturk and Aksoy [61] proposed a closed-loop supply chain optimization with a linear physical programming approach to maximize total profit and recovery rate and minimize the total recovery time. Golpira and Javanmardan [23] developed a bi-level multi-objective risk-based optimization model for a closed-loop supply chain to a decentralized decision system. A fuzzy multi-objective mixed integer linear programming model was proposed by Cinar [9], which aims to minimize the total cost of reverse logistics network and total carbon emissions for using EoL. Jabarzadeh et al. [32] recommended a multi-objective mixed integer linear model for sustainable fruit closed-loop supply chain network that maximizes profit, customer sentiment, and quality. Fahimnia et al. [19] and Nayeri et al. [55] designed optimization approaches for sustainable closed-loop supply chain networks. These models aim to minimize total cost, control carbon emissions, and maximize social impact. Specifically, Nayeri et al. [55] investigated the effect of carbon pricing in the textile industry, focusing on reducing both total cost and carbon emissions. Soleimani [78] developed a multi-objective model to minimize energy consumption and maximize profit for a sustainable closed-loop supply chain model for the mining industry. A robust fuzzy optimization model with the objectives of minimizing total cost and total carbon dioxide emission for a closed-loop supply chain network design problem in the electronics industry was proposed by Talaei et al. [82]. Lu et al. [48] designed an optimization model to minimize total cost and pollution by heavy metals of a green network considering heavy metal pollutants in the electronics industry. Rohmer et al. [69] proposed a multi-objective analysis (minimizing total cost and total environmental impact) for sustainable supply chain design in the food system. Tosarkani and Amin [86] proposed an environmental multi-objective optimization model (maximizing profit and environmental compliance) to configure a hybrid supply chain network in the battery industry.

Studies in the literature propose linear mathematical models, including EoL vehicles and reverse supply chains in terms of CE and sustainability in the automotive industry, and perform optimization studies. The objective function of some of these studies is maximization [59, 76, 77, 85]. In some studies, the objective function is minimization [12, 38, 62] . In some studies in the literature dealing with the automotive industry, the authors have developed a multi-objective linear model [45, 66]. A study also suggested a multi-objective nonlinear model in the automotive industry. Shahedi et al. [75] and Ozceylan et al. [59] proposed an LP model for a closed-loop supply chain network design with the objective of maximum revenue. A decision model for a strategic closed-loop supply chain to reclaim EoL Vehicles was proposed by Shankar et al. [76]. Simic [77] proposed a stochastic mixed integer linear programming model for managing end-of-life (EoL) vehicles to maximize expected profit. Torgul and Paksoy [85] proposed a multilevel competitive closed-loop supply chain network design and developed a mixed integer nonlinear programming to maximize total profit. A mixed integer linear programming model aimed to minimize the total cost of reverse logistics activities of EOL vehicles was developed by Demirel et al. [12]. A mixed integer linear programming for optimizing the fuzzy reverse supply chain for EoL vehicles was proposed by Phuc et al. [62]. Karagoz et al. [38] proposed a novel stochastic optimization model to minimize the total cost for reverse logistics network design of EoL vehicles. Kusakci et al. [45] aimed to minimize total cost and maximize the aspiration level of fuzzy constraints while optimizing the reverse logistics network of EoL Vehicles under fuzzy supply. Multi-objective mixed integer linear optimization model aimed at minimizing environmental effects and maximizing net profit for a sustainable closed-loop supply chain network was developed by Rajak et al. [66]. A stochastic multi-objective mixed integer nonlinear programming model (maximizing total profit, minimizing effects of pollutants, and maximizing the number of employment opportunities) for a sustainable closed-loop supply chain network design was developed by Shahedi et al. [75].

In Table 5.2, some important studies that have conducted OR approaches on sustainability and CE in the manufacturing sector are summarized.

The relationship between CE principles and OR approaches is symbiotic, with OR acting as a crucial enabler for the effective implementation of circular practices. As organizations worldwide strive to reduce their environmental footprint and transition toward more sustainable business models, the synergy between CE strategies and OR techniques will become increasingly essential. By harnessing the power of OR, companies can navigate the complexities of the CE with precision, ultimately driving environmental sustainability and economic efficiency hand in hand.

5.3 Results and Discussion

Today, customers do not evaluate the value of a company or a product based on its production process efficiency and low price. At the same time, customers evaluate a product's value according to the importance given to sustainability by the company that produces the product. CE, a sustainable development method, is an economic model that aims to maximize resource use efficiency by ensuring that materials and products circulate effectively and efficiently in biological and technical cycles.

Table 5.2 Circular economy and future trends

Future trend	Reference	Objective	Methodology	Industry of model application
Artificial intelligence	Weichhart et al. [93]	Designing intelligent enterprise systems	AI	Manufacturing (general)
	Abou Bakar et al. [1]	Automated AI-enhanced smartphone e-waste classification in smart cities	Machine learning	Smartphone
	Kumar et al. [46]	Sorting COVID-related medical waste streams from other waste types	Machine learning	Waste management
	Wilts et al. [96]	Sorting of municipal waste	Robotic sorting system	Municipal waste
	Andono et al. [2]	Optimization model of end-to-end value chain of onions	AI and IoT	Agricultural
	Pregowska et al. [64]	AI-based battery waste management	Machine learning	Electronics (battery)
	Neto et al. [11]	Economic and environmental optimization of the reverse network of waste	AI	Electrical and electronic
	Onyeaka et al. [56]	AI to tackle food waste and enhance the CE	AI	Food
Industry 4.0	Visich et al. [92]	Implementation of RFID systems in closed-loop supply chains	RFID	Manufacturing (general)
	Baccarelli et al. [4]	Energy-efficient networked computing architectures	Cloud Computing, Big Data and IoT	Manufacturing (general)
	Kim et al. [41]	A profitable and eco-friendly closed-loop supply chain system	RFID	Automotive remanufacturing
	Moreno et al. [52]	Re-distributed manufacturing and a product-service system approach	RdM and 3D printing	Shoe manufacturing

(continued)

Table 5.2 (continued)

Future trend	Reference	Objective	Methodology	Industry of model application
	Jabbour et al. [33]	Integration of the CE and big data	Big data	Manufacturing (general)
	Mastos et al. [50]	An IoT solution on the sustainable performance	IoT	Lift (scrap metal)
	Chauhan et al. [7]	Smart healthcare waste disposal system infused by CE aspects	DEMATEL method	Healthcare waste
	Andono et al. [2]	Optimization model of end-to-end value chain of onions	AI and IoT	Agricultural
	Tavana et al. [83]	Location inventory routing in an circular supply chain network	Lagrangian relaxation algorithm and Artificial IOT	Manufacturing (general)
Operational research models	Pishvaee et al. [63]	Min total cost	MILP	Manufacturing (general)
	Fahimnia et al. [19]	Min total cost and min carbon emission	MOMILP	Manufacturing (textile)
	Ene and Ozturk [16]	Max total profit of reverse supply chain network	MILP	Manufacturing (general)
	Ozceylan et al. [60]	Min total cost	MINLP	Electrical and electronics
	Demirel et al. [12]	Min total cost	MILP	Manufacturing (vehicles)
	Ozceylan et al. [59]	Max revenue	LP	Manufacturing (vehicles)
	Phuc et al. [62]	Min total cost	MILP	Manufacturing (vehicles)
	Shankar et al. [76]	Max contribution margin or profit	MILP	Automotive (general)
	Wu et al. [98]	Min total cost	MILP	Manufacturing (general)
	Hajipour et al. [27]	Max total profit	MINLP	Network Marketing
	Kir et al. [42]	Min total cost	MINLP	Recycling (tire)
	Kusakci et al. [45]	Min total cost and max aspiration level of fuzzy constraints	MOMILP	Manufacturing (vehicles)

(continued)

Table 5.2 (continued)

Future trend	Reference	Objective	Methodology	Industry of model application
	Rohmer et al. [69]	Min total cost and min total environmental impact	MOMILP	Food (nutrient)
	Simic [77]	Max expected profit	MILP	Manufacturing (vehicles)
	Tosarkani and Amin [86]	Max profit and max the environmental compliance	MOO	Battery
	Yazgan et al. [99]	Min transportation costs and cost of CO2 emissions	MINLP	Tire
	Jabarzadeh et al. [32]	Max profit, customer responsiveness and quality	MILP	Agriculture (fruit)
	Nayeri et al. [55]	Min the total cost, reduce and control total carbon emissions and max social impacts	MOMILP	Agriculture (water tank)
	Golpira and Javarmar-dan [23]	Min cost, max job opportunities and min work-related damages	MOMILP	Public and private sector
	Lu et al. [48]	Min total cost and min pollution by heavy metals	MOMILP	Electronics (copper)
	Ozturk and Aksoy [61]	Max total profit, max recovery rate and min total recovery time	MOMILP	Manufacturing (general)
	Soleimani [78]	Min energy consumption and max profit	MOO	Mining
	Torgul et al. [84]	Max total profit	MILP	Food (juice)
	Yildizbasi and Arioz [100]	Max total profit from the flow in the supply chain network	MILP	Electrical and electronics
	Cinar [9]	Min total reverse logistics network cost and min total carbon emissions	MILP	Energy (wind)
	Kabadurmus et al. [37]	Max total distributed food waste and min total cost	MILP	Food (general)

(continued)

Table 5.2 (continued)

Future trend	Reference	Objective	Methodology	Industry of model application
	Karagoz et al. [38]	Min total cost	MILP	Manufacturing (vehicles)
	Kazancoglu et al. [40]	Min total cost, CO_2 emissions, and particulate matter concentrations	MILP	Home appliances
	Rajak et al. [66]	Min environmental effects and max net profit	MOMILP	Automotive (steering column)
	Shahedi et al. [75]	Max total profit, min effects of pollutants, and max number of employment opportunities	MOMILP	Automotive (general)
	Torgul and Paksoy [85]	Max total profit	MINLP	Automotive (steel)

This chapter explains the beginning and the concept of the CE in the literature, and the differences of CE between the LE are stated. Studies conducted in the EU in line with Sustainability and CE are overviewed. The structure and principles of the CE are stated, and key studies on the benefits and barriers to implementing this concept are included. In addition, the benefits and barriers of the CE are listed. Classification and explanations of CE business models (models aimed at slowing and closing cycles) are given, which includes a review of the relationship between CE and product life cycle. Finally, this study explores the use of AI in integrating the CE into product life cycle processes.

Applying the CE concept to systems can solve variability in raw material costs in line with the LE concept. This system, which aims to reach the optimum level of resource use, covers the efficient use of resources at the optimum level and the conversion of wastes into new resources through recycling or repair. In this sense, it not only affects a product's life cycle but also provides interactive resource use between products.

In this chapter, to investigate the impact of future trends on CE, three trends were selected and examined: AI, Industry 4.0, and OR approaches. The study showed that AI, Industry 4.0, and OR approaches could play a crucial role in promoting the frequent use of new processes and products developed in line with the innovative and CE. Overall, this study shows that, through these approaches that have revolutionized the Manufacturing Industry, the stated barriers to the transition to the CE can be removed or reduced. By using the business models outlined, using the rising trends, and implementing CE strategies, the stated benefits of the CE can begin to be achieved. This chapter guides the implementation of the CE concept and provides a great contribution to the literature from many perspectives.

The impact of the CE on production processes is inevitable. To realize this concept, the structure of the processes in the product life cycle needs to be changed/improved. Future studies may contribute to the literature by making changes/improvements in processes to adopt the CE to a product life cycle. Research can be expanded with studies on the recognition and effectiveness of CE by companies and individuals. Strategies for the transition to the CE and the steps to be taken to implement the CE in the product life cycle can be determined, and the concept of the CE can be examined from the perspective of potential customers and employees. A real-life production model of the CE can be developed. Using CE and (separately or together) AI, Industry 4.0, and OR approaches in the production sector can help companies reduce waste, save resources, increase efficiency, and create new business opportunities. Thus, a real-life model can be developed for the CE and to adapt AI to the product life cycle.

Competing Interests The authors have no conflicts of interest to declare that are relevant to the content of this chapter.

References

1. Abou Baker N, Szabo-Müller P, Handmann U (2021) Transfer learning-based method for automated e-waste recycling in smart cities. EAI endorsed transactions on smart cities 5(16): e1-e1
2. Andono PN, Saputra FO, Shidik GF, Hasibuan Z A (2022, September). End-to-end circular economy in onion farming with the application of artificial intelligence and internet of things. In 2022 International seminar on application for technology of information and communication (iSemantic, Semarang, Indonesia.)
3. Andrews D (2015) The Circular Economy, Design Thinking and Education For Sustainability. Local Economy 30(3):305–315
4. Baccarelli E, Naranjo PGV, Scarpiniti M, Shojafar M, Abawajy JH (2017) Fog of everything: Energy-efficient networked computing architectures, research challenges, and a case study. 5:9882–9910
5. Bocken NMP, de Pauw I, Bakker C, van der Grinten B (2016) Product design and business model strategies for a circular economy. J Ind Prod Eng 33(5):308–320
6. Buruzs A, Torma A (2017) A Review on the Outlook of the Circular Economy in the Automotive Industry. International Journal of Environmental and Ecological Engineering 11(6):576–580
7. Chauhan A, Jakhar SK, Chauhan C (2021) The interplay of circular economy with industry 4.0 enabled smart city drivers of healthcare waste disposal. J Clean Prod 279:123854
8. Chauhan C, Parida V, Dhir A (2022) Linking circular economy and digitalisation technologies: A systematic literature review of past achievements and future promises. Technol Forecast Soc Change 177(3):121508
9. Cinar S (2021) SUSTAINABLE REVERSE LOGISTIC NETWORK DESIGN FOR END-OF-LIFE USE-CASE STUDY. RAIRO-Oper. Res. 55 (2021) S503–S521 RAIRO Operations Research https://doi.org/10.1051/ro/2019069

10. Cowls J, Tsamados A, Taddeo M, Floridi L (2023) The AI gambit: leveraging artificial intelligence to combat climate change–opportunities, challenges, and recommendations. AI Soc 38:283–307
11. de Oliveira Neto GC, de Araujo SA, Alliprandini DH, Flausino FR, Amorim M (2023) Simulation of electronic waste reverse chain for São Paulo circular economy: An artificial intelligence-based approach for economic and environmental optimization. Sensors 23(22):9046
12. Demirel E, Demirel N, Gökçen H (2016) A mixed integer linear programming model to optimize reverse logistics activities of end-of-life vehicles in Turkey. Journal of Cleaner Production 112:2101–2113
13. Ellen MacArthur Foundation (2013) Towards the circular economy Vol. 1: an economic and business rationale for an accelerated transition. Available via Ellen MacArthur Foundation. https://ellenmacarthurfoundation.org/towards-the-circular-economy-vol-/1-an-economic-and-business-rationale-for-an Cited 25 Jul 2023
14. Ellen MacArthur Foundation and McKinsey Center for Business and Environment (2015) Growth Within: A Circular Economy Vision for a Competitive Europe. Available via Ellen MacArthur Foundation. https://ellenmacarthurfoundation.org/growth-within-a-circular-economy-/vision-for-a-competitive-europe Cited 25 Jul 2023
15. Ellen MacArthur Foundation (2022) The butterfly diagram: Visualising the circular economy. Retrieved from https://ellenmacarthurfoundation.org/circular-economy-diagram. Cited 09 April 2023
16. Ene S, and Öztürk N (2015) Network modeling for reverse flows of end-of-life vehicles, Waste Manag., 38 , pp. 284–296
17. Esposito M, Tse T, Soufani K (2015) Is the Circular Economy a New Fast Expanding Market? Thunderbird International Business Review 59(1):9–14
18. European Parliament and of the Council. (2000). Directive 2000/53/EC of the European Parliament and of the Council of 18 September 2000 on end-of life vehicles. Available via EUMonitor. https://www.eumonitor.eu/9353000/1/j9vvik7m1c3gyxp/vhckn6df9zzn Cited 11 Nov 2023
19. Fahimnia B, Sarkis J, Dehghanian F, Banihashemi N, Rahman S (2013) The impact of carbon pricing on a closed-loop supply chain: an Australian case study. Journal of Cleaner Production 59:210–225
20. Fetting C (2020) The European Green Deal, ESDN Report, ESDN Office, Vienna.
21. Gedik Y (2020) Understanding the circular economy: A theoretical framework. Turk Bus J 1(2):13–40
22. Ghoreishi M, Happonen A (2019) New Promises AI Brings into Circular Economy Accelerated Product Design: Review on Supporting Literature. In: 7th International Conference on Environment Pollution and Prevention, Melbourne, Australia.
23. Golpira H, Javanmardan A (2021) Decentralized decision system for closed-loop supply chain: a bi-level multi-objective risk-based robust optimization approach. Computers & Chemical Engineering 154:107472
24. Grafström J, Aasma S (2021) Breaking circular economy barriers. Journal of Cleaner Production 292:126002
25. Gupta PK, Shree V, Hiremath L, Rajendran S (2019) The use of modern technology in smart waste management and recycling: artificial intelligence and machine learning. Recent Adv Comput Intell 823:173–188
26. Haas W, Krausmann F, Wiedenhofer D, Heinz M (2015) How Circular is the Global Economy?: An Assessment of Material Flows, Waste Production, and Recycling in the European Union and the World in 2005. Journal of Industrial Ecology 19(5):765–777
27. Hajipour V, Tavana M, Di Caprio D, Akhgar M, Jabbari Y (2019) An optimization model for traceable closed-loop supply chain networks. Applied mathematical modelling 71:673–699
28. Hart J, Adams K, Giesekam J, Tingley DD, Pomponi F (2019) Barriers and drivers in a circular economy: the case of the built environment. Procedia CIRP 80:619–624

29. Homrich AS, Galvão G, Abadia LG, Carvalho MM (2018). The circular economy umbrella: Trends and gaps on integrating pathways. Journal of Cleaner Production, 175:525–543.
30. Hopkinson P, Chen HM, Zhou K, Wong Y, Lam D (2018) Recovery and Re-Use of Structural Products from End of Life Buildings. Proceedings of the Institution of Civil Engineers - Engineering Sustainability 172(3):1–36
31. İşler G, Akyol DE, Yazgan HR (2023) Theory and research concerning the circular economy model and future trend. In: International Symposium on Intelligent Manufacturing and Service Systems. Springer Nature Singapore, pp 769–782
32. Jabarzadeh Y, Reyhani Yamchi H, Kumar V, Ghaffarinasab N (2020) A multi-objective mixed-integer linear model for sustainable fruit closed-loop supply chain network. Management of Environmental Quality: An International Journal 31(5):1351–1373
33. Jabbour CJC, de Sousa Jabbour ABL, Sarkis J, Godinho Filho M (2019) Unlocking the circular economy through new business models based on large-scale data: an integrative framework and research agenda. Technol Forecast Soc Change 144:546–552
34. Jesus A, Mendonça S (2018) Lost in transition? Drivers and barriers in the eco-innovation road to the circular economy. Ecological Economics 145:75–89
35. Jawahir IS, Bradley R (2016) Technological elements of circular economy and the principles of 6R-based closed-loop material flow in sustainable manufacturing. Procedia CIRP, 40:103–108.
36. John Lekan A, Oloruntoba AS (2019) Artificial intelligence in the transition to circular economy. American Journal of Engineering Research 9(6):185–190
37. Kabadurmuş O, Kazançoğlu Y, Yüksel D, Pala MÖ (2022) A circular food supply chain network model to reduce food waste. Annals of Operations Research 1–31
38. Karagöz S, Aydön N, Simic V (2022) A novel stochastic optimization model for reverse logistics network design of end-of-life vehicles: A case study of Istanbul. Environmental Modeling & Assessment 27(4):599–619
39. Kaya A, Alumur Alev S (2017) Closed Loop Supply Chain Network Design. Industrial Engineering 28(2):2–18
40. Kazançoğlu Y, Yüksel D, Sezer MD, Mangla SK, Hua L (2022) A green dual-channel closed-loop supply chain network design model. Journal of Cleaner Production 332(1):130062
41. Kim YW, Chang TW, Park J (2017) Gen2 RFID-based system framework for resource circulation in closed-loop supply chains. Sustainability 9(11):1995
42. Kör S, Ercan Cömert S, Yener F, Yazgan HR, Candan G (2019) Hazardous waste recycling: end of life tires case. Polish Acad Sciences Inst Physics 135(4):681–683
43. Kirchherr J, Piscicelli L, Bour R, Kostense-Smit E, Muller J, Huibrechtse-Truijens A et al. (2018) Barriers to the circular economy: Evidence from the European Union (EU). Ecological Economics 150:264–272
44. Korhonen J, Honkasalo A, Seppälä J (2018) Circular Economy: The Concept and its Limitations. Ecological Economics 143:37–46
45. Kuşakçı AO, Ayvaz B, Cin E, Aydön N (2019) Optimization of reverse logistics network of End of Life Vehicles under fuzzy supply: A case study for Istanbul Metropolitan Area. Journal of Cleaner Production 215:1036–1051
46. Kumar NM, Mohammed MA, Abdulkareem KH, Damasevicius R, Mostafa SA, Maashi MS, Chopra SS (2021) Artificial intelligence-based solution for sorting COVID related medical waste streams and supporting data-driven decisions for smart circular economy practice. Process Saf Environ Prot 152:482–494
47. Lacy P, Long J, Spindler W (2020) The Circular Economy Handbook: Realizing the Circular Advantage. Palgrave Macmillan.
48. Lu S, Hou W, Li Z, Xia J, Xie L, Su H (2022) Design and optimisation of a green manufacturing-recycling network considering heavy metal pollutants–an electronic assembly case. International Journal of Production Research 60(9):2830–2849
49. Lyle JT (1994) Regenerative design for sustainable development, Rethinking the Mind in Nature. John Wiley.
50. Mastos TD, Nizamis A, Vafeiadis T, Alexopoulos N, Ntinas C, Gkortzis D, Papadopoulos A, Ioannidis D, Tzovaras D (2020) Industry 4.0 sustainable supply chains: An application of an IoT enabled scrap metal management solution. J Clean Prod 269:122377

51. McDonough W, Braungart M, Clinton B (2013) The upcycle: Beyond sustainability-designing for abundance. North Point Press.
52. Moreno M, Turner C, Tiwari A, Hutabarat W, Charnley F, Widjaja D, Mondini L (2017) Re-distributed manufacturing to achieve a circular economy: A case study utilizing IDEF0 modeling. Procedia CIRP 63:686–691
53. Murray A, Skene K, Haynes K (2017) The Circular Economy: An interdisciplinary exploration of the concept and its application in a global context. Journal of Business Ethics 140(3):369–380
54. Nañez Alonso SL, Reier Forradellas RF, Pi Morell O, Jorge Vazquez J (2021) Digitalization, circular economy and environmental sustainability: The application of artificial intelligence in the efficient self-management of waste. Sustainability 13(4):2092
55. Nayeri S, Paydar MM, Asadi-Gangraj E, Emami S (2020) Multi-objective fuzzy robust optimization approach to sustainable closed-loop supply chain network design. Computers and Industrial Engineering 148:106716
56. Onyeaka H, Tamasiga P, Nwauzoma UM, Miri T, Juliet UC, Nwaiwu O, Akinsemolu AA (2023) Using artificial intelligence to tackle food waste and enhance the circular economy: Maximising resource efficiency and minimising environmental impact: A review. Sustainability 15(13):10482
57. Ortego A, Calvo G, Valero A, Iglesias-Émbil M, Valero A, Villacampa M (2020) Assessment of strategic raw materials in the automobile sector. Resour Conserv Recycl 161:104968
58. Özceylan E (2016) Simultaneous optimization of closed-and open-loop supply chain networks with common components. Journal of Manufacturing Systems 41:143–156
59. Özceylan E, Demirel N, Çetinkaya C, Demirel E (2017) A closed-loop supply chain network design for automotive industry in Turkey. Computers and Industrial Engineering 113:727–745
60. Özceylan E, Paksoy T, Bektaş T (2014) Modeling and optimizing the integrated problem of closed-loop supply chain network design and disassembly line balancing. Transportation research part E: logistics and transportation review 61:142–164
61. Öztürk G, Aksoy HK (2021) Closed Loop Supply Chain Optimization with Linear Physical Programming Approach. Journal of Statistics and Applied Sciences 2(2):14–28
62. Phuc PNK, Vincent FY, Tsao YC (2017) Optimizing fuzzy reverse supply chain for end-of-life vehicles. Computers and Industrial Engineering 113:757–765
63. Pishvaee MS, Rabbani M, Torabi SA (2011) A robust optimization approach to closed-loop supply chain network design under uncertainty. Applied mathematical modelling 35(2):637–649
64. Pregowska A, Osial M, Urbańska W (2022) The application of artificial intelligence in the effective battery life cycle in the closed circular economy model—A perspective. Recycling 7(6):81
65. Preston F (2012) A global redesign? Shaping the circular economy briefing paper. Chatham House.
66. Rajak S, Vimal KEK, Arumugam S, Parthiban J, Sivaraman SK, Kandasamy J et al (2022) Multi-objective mixed-integer linear optimization model for sustainable closed-loop supply chain network: A case study on remanufacturing steering column. Environment, Development and Sustainability 24(5):6481–6507
67. Ritzén S, Sandström GÖ (2017) Barriers to the Circular Economy – Integration of Perspectives and Domains. Procedia CIRP 64:7–12
68. Roberts H, Zhang J, Bariach B, Cowls J, Gilburt B, Juneja P, Tsamados A, Ziosi M, Taddeo M, Floridi L (2022) Artificial intelligence in support of the circular economy: Ethical considerations and a path forward. AI Soc 1:1–14
69. Rohmer SUK, Gerdessen JC, Claassen GDH (2019) Sustainable supply chain design in the food system with dietary considerations: A multi-objective analysis. European Journal of Operational Research 273(3):1149–1164
70. Rojek I, Jasiulewicz-Kaczmarek M, Piechowski M, Miko lajewski D (2023) An artificial intelligence approach for improving maintenance to supervise machine failures and support their repair. Appl Sci 13(8):4971

71. Sapmaz Veral E (2021) Döngüsel Ekonomi: Engeller, Stratejiler ve İş Modelleri. Ankara Üniversitesi Çevrebilimleri Journal 8(1):7–18
72. Sapmaz Veral E, Yiğitbaşıoğlu H (2018) Avrupa Birliği atök politikasönda atök yönetiminden kaynak yönetimi yaklaşımına geçiş yönelimleri ve döngüsel ekonomi modeli. Ankara Üniversitesi Çevrebilimleri Journal 6(1):1–19
73. Schempp C, Hirsch P (2020) Categorisation system for the circular economy – A sector-agnostic categorisation system for activities substantially contributing to the circular economy. Publications Office of the European Union, Brussels, Belgium.
74. Sehnem S, Vazquez-Brust D, Pereira SCF, Campos LMS (2019) Circular Economy: Benefits, impacts and overlapping. Supply Chain Manage Int J 24(6):784–804
75. Shahedi A, Nasiri MM, Sangari MS, Werner F, Jolai F (2022) A stochastic multi-objective model for a sustainable closed-loop supply chain network design in the automotive industry. Process Integration and Optimization for Sustainability 6(1):189–209
76. Shankar R, Bhattacharyya S, Choudhary A (2018) A decision model for a strategic closed-loop supply chain to reclaim End-of-Life Vehicles. International Journal of Production Economics 195:273–286
77. Simic V (2019) Interval-parameter conditional value-at-risk two-stage stochastic programming model for management of end-of-life vehicles. Environmental Modeling and Assessment 24(5):547–567
78. Soleimani H (2021) A new sustainable closed-loop supply chain model for mining industry considering fixed-charged transportation: A case study in a travertine quarry. Resources Policy 74:101230
79. Stahel WR (2016) The circular economy. Nature 531(7595):435–438
80. Stahel WR, Reday-Mulvey G (1976) The potential for substituting manpower for energy: A report to the European Commission. Vantage Press.
81. Stahel WR, Reday-Mulvey G (1977) The Potential for Substituting Manpower for Energy: Final Report 30 July 1977 for the Commission of the European Communities. Battelle, Geneva Research Centre.
82. Talaei M, Moghaddam BF, Pishvaee MS, Bozorgi-Amiri A, Gholamnejad S (2016) A robust fuzzy optimization model for carbon-efficient closed-loop supply chain network design problem: a numerical illustration in electronics industry. Journal of cleaner production 113:662–673
83. Tavana M, Khalili Nasr A, Santos-Arteaga FJ, Saberi E, Mina H (2023) An optimization model with a Lagrangian relaxation algorithm for artificial Internet of Things-enabled sustainable circular supply chain networks. Ann Oper Res:1–36
84. Torgul B, Demiralay E, Paksoy T (2021) Network Design and Optimization in the Agri-Food Supply Chain: A Juice Business Example. Journal of Transportation and Logistics 6(1):49–70
85. Torgul B, Paksoy T (2023) Multi-level competitive closed loop supply chain network design. Journal of Polytechnic.
86. Tosarkani BM, Amin SH (2019) An environmental optimization model to configure a hybrid forward and reverse supply chain network under uncertainty. Computers and Chemical Engineering 121:540–555
87. Turner RK, Pearce DW (1990). The ethical foundations of sustainable economic development. London, UK: International Institute for Environment and Development.
88. Türkiye Foundation for Combating Erosion, Afforestation and Protection of Natural Assets (TEMA Foundation) (2015) We will need three worlds by 2050. Retrieved from https://www.tema.org.tr/basin-odasi/basin-bultenleri/2050de-uc-tane-dunyaya-ihtiyacimiz-olacak. Cited 11 November 2023
89. United Nations Environment Programme (2021) The Role of Business in Moving from Linear to Circular Economies. Retrieved from https://wedocs.unep.org/bitstream/handle/20.500.11822/36830/RBMLCE.pdf. Cited 25 April 2022
90. United Nations Member States (2015) United Nations Sustainable Development Goals. Retrieved from https://sdgs.un.org/goals. Cited 25 April 2022
91. Ünlütürk Ş, Kihtir Öztürk P, Kardeş S, Birdal M, Saral B, Aşıroğlu B et al. (n.d.). Circular economy guide for businesses. https://business4goals.org/. Cited 25 Jul 2023

92. Visich JK, Li S, Khumawala BM (2007) Enhancing product recovery value in closed-loop supply chains with RFID. J Manager Issues:436–452
93. Weichhart G, Molina A, Chen D, Whitman L, Vernadat F (2016) Challenges and current developments for Sensing, Smart and Sustainable Enterprise Systems. Comput Ind 79:34–46
94. Wijkman A, Skånberg K (2015) The Circular Economy and Benefits for Society: Jobs and Climate Clear Winners in an Economy Based on Renewable Energy and Resource Efficiency. Club of Rome.
95. Wilson M, Paschen J, Pitt L (2022) A Circular Economy Meets Artificial Intelligence (AI): Understanding the Opportunities of AI for Reverse Logistics. Manag Environ Qual Int J 33(1):9–25
96. Wilts H, Garcia BR, Garlito RG, Gómez LS, Prieto EG (2021) Artificial intelligence in the sorting of municipal waste as an enabler of the circular economy. Resources 10(4):28
97. Wise C, Pawlyn M, Braungart M (2013) Eco-engineering: Living in a materials world, Nature 494, 172–175
98. Wu GH, Chang CK, Hsu LM (2018) Comparisons of interactive fuzzy programming approaches for closed-loop supply chain network design under uncertainty. Computers & Industrial Engineering 125:500–513.
99. Yazgan HR, Kör S, Yener F, Comert SE (2020) A New Collecting and Management Proposal Under Logistics 4.0 and Green Concept. In: Logistics 4.0, CRC Press, 320–337.
100. Yöldözbaşı A, Arööz Y (2021) A novel hybrid network optimization model for printed circuit boards recycling: a circular economy perspective. Clean Technologies and Environmental Policy 23:2989–3013
101. Yuan Z, Bi J, Moriguichi Y (2006) The Circular Economy: A New Development Strategy in China, Journal of Industrial Ecology 10(1–2):4–8

Chapter 6
Sales Center Location, Green Open Vehicle Routing for Product Transport in the Urban Context and Their Future Perspectives with the Circular Economy

José Rodríguez-Melquiades and Lourdes Ramirez Cerna

Abstract Cities in various countries, including those in South America, are undergoing changes to improve living conditions. One such change involves acquiring products efficiently and promptly. This is the current reality of the globalized world. In this context of a consumerist society, locating sales centers and vehicle routing constitute a great support to optimize the process of serve people. This proposal develops an optimization model based on the mixed integer linear programming methodology and a new hybrid algorithm based on the k-means algorithm and local search. Therefore, they are a great contribution to the supply chain and environmental care because they not only optimize the process of serving customers by applying location-routing problems but also determine fuel consumption and measure the CO_2 generation. The model implemented in GLPK (GNU Linear Programming Kit) obtains optimal results in small scenarios of up to 19 customers. For larger scenarios of 30 or 50 customers, the hybrid algorithm obtains good results in short processing times.

Keywords Location · Green open vehicle routing · Product transport · Urban context · Circular economy

J. Rodríguez-Melquiades (✉)
Optimización en logística y combinatoria - OPLOCOM, Departamento de Informática, Universidad Nacional de Trujillo, Trujillo, Peru
e-mail: jrodriguez@unitru.edu.pe

L. R. Cerna
Optimización en logística y combinatoria - OPLOCOM, Carrera de Ingeniería de Sistemas, Universidad de Lima, Lima, Peru
e-mail: lramirec@ulima.edu.pe

B. Jarboui et al. (eds.), *Circular Economy Supply Chains*, Computational Intelligence Methods and Applications,
https://doi.org/10.1007/978-981-96-7905-8_6

6.1 Introduction

Everyday life in modern society has undergone major changes of various kinds. This is due to the fact that in the globalized world, people show greater demand for acquiring products of various types for their consumption and therefore satisfaction. The need that is being created has influenced technological advancement, which is obtained by sales centers to enable them to provide accurate and prompt customer service.

Due to globalization, people involved in the production process and other activities are often too busy to visit sales centers, in both developed and developing countries. Therefore, it is necessary to locate a sales center that will act as a facility, which will provide a door-to-door service to the so-called customers through a vehicle routing process, i.e., a logistics city.

Given this reality, global consumption generates various types of waste, including solid urban, agricultural, construction, industrial, mining, and medicinal wastes, which, if not properly collected and/or treated, pollute the environment, because according to [13] on average each person generates 750 gr/da$y$ of waste which translates into more than two billion tons of waste annually worldwide. Given this situation, the concept and implementation of a circular economy are aimed at protecting our planet, since it is necessary to build a sustainable society through the development of such an economy [36]. When waste can be reused or recycled, it returns to the productive cycle. In addition, environmental awareness of citizens is necessary, regardless of the country they belong to, since any type of contribution is welcome.

Therefore, it is necessary to develop optimization models, metaheuristics, and hybrid algorithms to meet the demands of transportation. At this point, there is a concern because transportation by vehicles is highly polluting, since they use fossil fuels and generate CO_2 per kg/ton-km traveled. This means that the transportation sector generates approximately 29% of total greenhouse gas emissions and 25% of global energy related to CO_2 emissions [20]. Ultimately there are proposals to use vehicles using alternative fuels, electric and hybrid vehicles [12].

Based on the state of the art on vehicle routing reviewed in [40], as well as its applications investigated in [33], our contribution presents a two-level optimization model and a new hybrid algorithm applied in the urban context, to serve customers who initially locate the sales centers for products such as groceries, home appliances, medicines, etc. Previously the suppliers supply the sales centers. At a later stage, the products will be transported to the customers using open vehicle routing, since it is considered that the transportation is outsourced to a specialized company with the purpose that the sales center minimizes maintenance costs of the vehicle fleet, besides saving on personnel costs. Therefore, they are a great contribution to the supply chain and environmental care because they not only optimize the process of serving customers but also determine fuel consumption and measure the CO_2 generation. See Fig. 6.1.

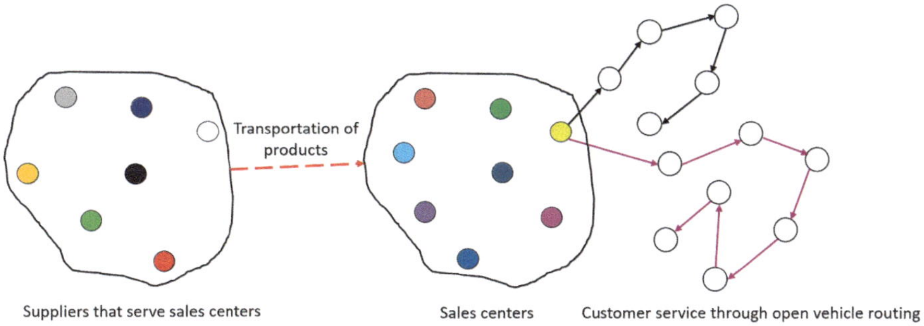

Fig. 6.1 Location of sales centers and open vehicle routing

The research presented is structured in the following sections: Sect. 6.2 describes the research method used to achieve the proposed objective. Section 6.3 consists of a review of the literature related to the proposed research. In Sect. 6.4 the problem is discussed by means of a description and formulation of an optimization model based on the mixed integer linear mixed programming methodology with two objective functions, in addition to a hybrid algorithm based on the k-means algorithm and local search. Section 6.5 shows the computational results and discussions. Section 6.6 analyzes the contribution of the circular economy in the urban context. Finally, Sect. 6.7 presents the conclusions of the research.

6.2 Research Method

Understanding current quantitative research in logistics, supported by combinatorial optimization and integration of location-routing and the circular economy concept, led to a search in the Scopus database to determine contributions in the urban context. The search focused on research from 2021 to 2023, using filters for review and research articles where "green logistics," "location-routing problem," and "circular economy" were keywords to gather information across different contexts.

The result was 153 articles with which relationships were found by means of the reserved words existing between them, i.e., a co-occurrence network was constructed with VOSViewer [9]. In addition, the complete count of all keywords with a minimum of two occurrences was considered. Figure 6.2 shows the relationship found between circular economy and sustainability.

According to Stevels [37], the circular economy prioritizes resources over emissions, assuming that the entire life cycle of materials is considered. Therefore, the objective is to maximize value creation and at the same time to reduce the use of resources, especially those offered by nature. Sustainability guarantees the quality

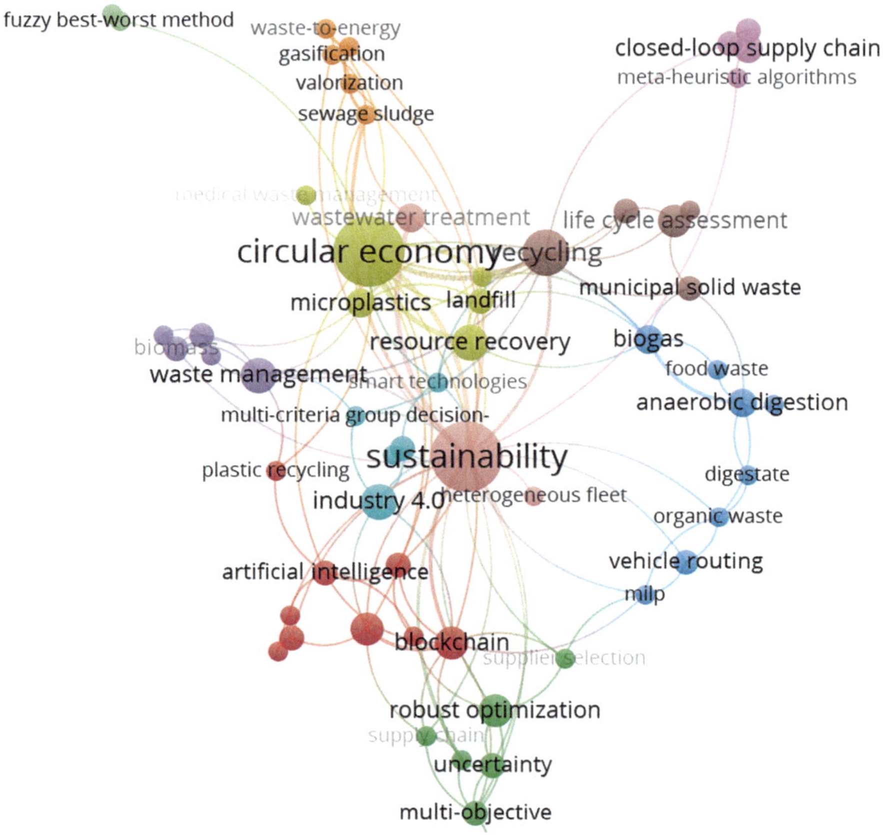

Fig. 6.2 Co-occurrence of keywords

of life in the urban space and balance in the social classes, thanks to the intersection between the economic, social, and environmental pillars.

In conclusion, urban sustainability is an important issue in modern society, being one of the most desired objectives to be achieved for the well-being of people through the use of available natural resources. Recently, the concept of "smart cities" has emerged, where smart technologies play an active role.

6.3 Review of Specialized Literature

In this section, we present a review of the literature as the foundation for the proposal made in this chapter.

6.3.1 City Logistics and Circular Economy

By 2023, the world population has increased at an annual growth rate of 0.91%, leading to various problems in cities and compelling authorities to make concrete decisions for citizen welfare. The world's population is expected to increase by nearly 2 billion persons in the next 30 years, from the current 8 billion to 9.7 billion in 2050, and could peak at almost 10.4 billion in the mid-2080s (https://www.un.org/en/global-issues/population).

Population growth has determined and will continue to bring migration of people from rural to urban areas within countries. According to the United Nations, 68% of the population will live in cities by 2050 [24]. This generates an imbalance in large cities, so it is necessary, for example, to take actions for the production and supply of various urban products. Logistics emerges as an alternative solution through the concept of a logistics city, which aims to optimize cargo distribution and improve efficiency in urban areas, mitigating congestion and environmental externalities.

To meet the high demand of consumers, logistics requires vehicles that use fuel to cover necessary travel distances. Unfortunately, the transportation in Canada, for example, is responsible for a 54% increase in greenhouse gas (GHG) emissions from 1990 to 2019 [3]. Fortunately, the market for electric vehicles is currently growing significantly and will serve the transportation sector well [3, 35].

As a result of the enormous activities that occur there, different goods are supplied through transportation units in logistics cities. Depending on its value, large quantities of waste produced can be recycled, rejected, or rejected. This will be possible if the concept of circular economy, a model of production and consumption, is put into practice [22].

6.3.2 Location-Routing Problem

Locating a distribution center, factory, or other type of facility is crucial as it initiates the supply process for urban population centers [30].

Bouchery et al. [4] defined green facility location as one where CO_2 emissions are minimized. This is possible through a variant of the location model based on the integer linear programming for capacitated facility location, as presented in [27] in whose work the amount of CO_2 generated by vehicles during customer service is determined.

Combinatorial optimization, particularly through integer linear programming, is a useful method for modeling computationally challenging problems, especially in large scenarios. A particular case of combinatorial problem is the vehicle routing problem—VRP with its different variations (different adaptations to real cases) discussed in [10], the objective being to find a set of optimal routes such that each customer can be visited by exactly one vehicle, which starts and ends its route at a depot.

While it is possible to obtain exact or optimal solutions quickly for moderate-sized scenarios using integer linear programming, it becomes challenging to do so in real-world situations with large datasets, as processing times become unacceptably long. Fortunately, the advancement of computation is contributing greatly to obtaining appropriate solutions (heuristics and metaheuristics) for large scenarios and in short computational processing time [40]. The metaheuristics are divided into local search and population search; many are inspired by many phenomena of life and nature [10].

An important variant for this research is open vehicle routing. In this case, the vehicle does not return to the depot at the end of its route. This approach is often used when the distribution company does not own its fleet of vehicles; therefore, returning the vehicle to the depot would unnecessarily increase economic and pollution costs.

An optimization model and an iterated local search algorithm for open vehicle routing with time windows were proposed by [5], whose computational results obtained using a set of 418 instances were considered good. Another proposal with capacity and distance constraints was presented in [29] whose solution was obtained using a genetic algorithm.

The authors in [18] discussed a case of capacitated open vehicle routing with time windows, where the objective function of the proposed optimization model considered vehicle travel time. Additionally, they proposed a greedy algorithm and a tabu search metaheuristic. The results showed that the metaheuristic outperformed other approaches. A recent application of this approach can be found in [28], as illustrated in Fig. 6.3.

The integration of the combinatorial problems of locating a set of suppliers and providing a service through optimal routes in an integrated way is being widely used, in particular, in logistics and supply chains. Given the importance of this type

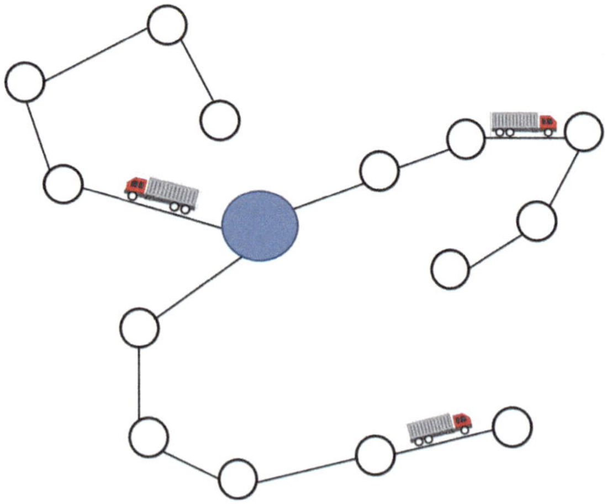

Fig. 6.3 Open vehicle routing with three routes

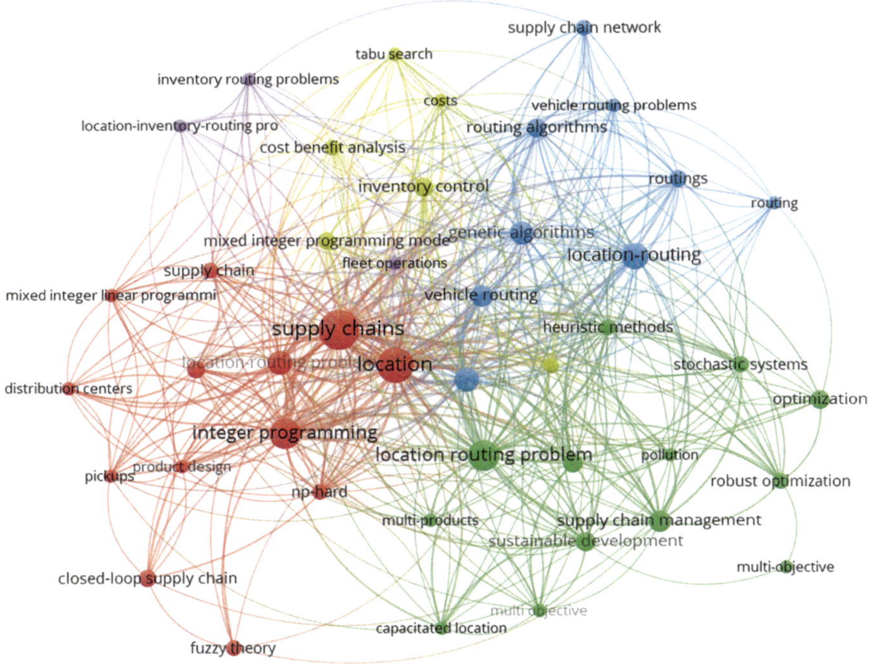

Fig. 6.4 Solution strategies for localization-routing

of problem and to know the current proposals and/or advances, respectively, this subsection presents a bibliometric analysis of this topic.

The research conducted in the Scopus database generated 54 articles closely related to the topic, for which "location-routing," "problem," "supply," "chain," and "sustainable" were used as search words. The result of VOSViewer was the identification of co-occurrence networks for which three occurrences were considered. Figure 6.4 shows the exact solution strategies using mixed integer linear programming and approximate strategies such as tabu search metaheuristics, genetic algorithms, particle swarm optimization, and other heuristic methods for multi-objective functions, as well as application cases in logistics.

Biomass is used to produce biofuel, bioenergy, biochemicals, and other biological products. Given the importance of this resource and to conserve the environment, an application for the biomass supply chain based on localization-routing was proposed by Cao et al. [6], using a mixed integer programming model and a tabu search metaheuristic.

A review of articles published in journals based on location-routing between the years 2014 and 2020 was analyzed in [39]; the contribution was to classify them by their bi-objective and multi-objective functions. The author used the insights gained to make suggestions for future research.

Table 6.1 is a comparative summary of the articles in Sect. 6.3 and other articles investigated. The similarities that exist between the articles are the use of vehicle routing and location-routing, where the ways to solve are through different optimization models and algorithmic strategies. Thus, the information obtained motivated the realization of a new proposal paper.

The circular economy is the business opportunity of our time. However, the world is currently only 8.6% circular. If there is no increase in circularity and if current consumption levels continue, the World Business Council for Sustainable Development calculated that the ecological resources of 2.3 planets will be needed by 2050. Circular economy supply chains, supported by circular economy logistics, offer a more sustainable path forward and offer businesses many benefits (https://www.maersk.com/insights/sustainability/2023/03/02/circular-economy-logistics).

This percentage is the reason why there is currently little research related to optimization, although recently the authors in [23] established in a systematic review that Reverse Logistics (RL) of end-of-use/end-of-life products has become a vital part of circular economy practices for manufacturers. Therefore, the future for the use of optimization is promising, as already discussed in Sect. 6.6.

6.4 Sales Center Location and Green Open Vehicle Routing for Product Transport

The section is constituted by an optimization model and a hybrid algorithm, which are a sustainable quantitative proposal in the urban context whose target is to find two objective functions.

6.4.1 Optimization Model

The proposed two-level model, based on mixed integer linear programming, is bi-objective and involves open routing. The first level consists of locating the suppliers by the sales centers. At this stage, only a supplier-sales center transportation problem is considered. In stage 2, sales centers, acting as depots, serve customers by providing delivery services based on a route plan obtained through open vehicle routing.

The type of routing considered assumes that the fleet does not belong to the sales centers because it is outsourced to another company that provides transportation services; therefore, once the last customer has been served, the vehicle returns to its depot, since the contract ends at this stage. Finally, as the process is environmentally friendly, at level 2 the amount of fuel used is measured, and consequently, the amount of CO_2 generated by the vehicles can be calculated.

Table 6.1 A comparative between the proposal paper and reviewed papers

Papers	VRP	Location-routing	Model	Algorithm	Fuel used	CO_2/GHG	Circular economy
Proposal	Green open	x	Bi-objective	Hybrid based on k-means and local search	Diesel	CO_2	x
[3]	Green		Bi-objective	ALNS	Diesel; electric	GHG	
[5]	Open		Objective	Iterated local search			
[7]	Green		Objective	Variable neighborhood search	Diesel; electric		
[14]		x	Bi-objective				x
[15]		x	Bi-objective	MOGWO; NSGA-II	Diesel	CO_2	
[16]		x	Three-objective	NSGA-II; MOSFS	Diesel	GHG	
[18]	Open		Objective	Greedy; tabu search			
[21]	Green		Objective		Diesel	CO_2	x
[22]	Green open	x	Bi-objective	Particle swarm optimization	Diesel	CO_2	x
[25]	x		Objective				
[26]		x	Objective	K-means clustering			
[27]	Green	x	Bi-objective		Diesel	CO_2	
[28]	Green open		Bi-objective	Genetic algorithm; local search	Diesel	CO_2	
[29]	Open		Objective	Genetic			
[32]		x	Bi-objective	Hybrid heuristic	Electric		x
[34]	Open	x		Branch-and-cut	Diesel		
[35]	Electric		Objective	Swarm intelligence	Electric		
[41]	Multiple trips		Objective		Diesel		x
[42]	Time windows		Objective	Hybrid with clustering and SGA-PSO			
[43]	Electric		Objective	Adaptive large neighborhood search	Electric		
[44]	x		Three-objective	Hybrid population optimization	Diesel		

Sets

S:	set of suppliers
J:	set of sales centers
I:	customer set
K:	set of vehicles for transporting products
$SJ := S \cup J$:	network at level 1
$P := J \cup I$:	network at level 2

Parameters

$d1_{sj}$:	distance between supplier $s \in SJ$ with sales center $j \in SJ$
$d2_{ij}$:	distance between sales center $i \in P$ with customers $j \in P$
t_j:	sales center capacity $j \in J$
$c_{ij} = freight \times \frac{d2_{ij}}{r}$:	transportation cost between $i \in P$ and $j \in P$
$freight$:	price to be paid for mobilizing customer orders
r:	fixed value per km traveled
f_j:	fixed cost to operate the sales center $j \in J$
dd_i:	customer demand $i \in I$
q_k:	vehicle capacity $k \in K$
tt_k:	maximum capacity of the distributor vehicle fuel tank k
cc:	vehicle fuel consumption in liters per km traveled
em_k:	CO_2 emission of the distributor vehicle in g/L

Variables

W_{sj}:	products transported from supplier $s \in S$ to sales center $j \in J$
$X_j \in \{0, 1\}$:	binary variable where 1 means that sales center $j \in J$ has been located, otherwise 0.
$Y_{ij} \in \{0, 1\}$:	binary variable whose value is interpreted as demand not served or served by sales center $j \in J$ to the customer $i \in I$
$Z_{ij} \in \{0, 1\}$:	binary variable for routing, where $Z_{ij} = 1$ means that every arc ij is the best route
U_{ij}:	quantifies the delivery of the product to the customer $i \in P$ before serving the next customer $j \in P$
$M_k \in \{0, 1\}$:	report whether or not a vehicle $k \in K$ is not in service
YY_{ik}:	amount of fuel in the vehicle $k \in K$ when it is $i \in P$ before reaching the next customer

Having established the abovementioned note, the proposed model is presented below:

$$Min : \sum_{s \in S, j \in J} d1_{sj} W_{sj} + \sum_{j \in J} f_j X_j + \sum_{i \in P, j \in I, k \in K : i \neq j} c_{ij} Z_{ijk} \tag{6.1}$$

$$Min : \sum_{i \in P, j \in I, k \in K : i \neq j} cc \left(\frac{em_k}{q_k} \right) \times d2_{ij} Z_{ijk} \tag{6.2}$$

s.t.

$$\sum_{s \in S} W_{sj} \leq t_j X_j, \forall j \in J \tag{6.3}$$

$$\sum_{s \in S} W_{sj} - \sum_{i \in I} dd_i Y_{ij} = 0, \forall j \in J \tag{6.4}$$

$$\sum_{i \in P: i \neq j} Z_{ijk} = 1, \forall j \in P, k \in K : j \neq \text{sales centers} \tag{6.5}$$

$$\sum_{j \in P: i \neq j} Z_{ijk} = 1, \forall i \in P, k \in K : i = \text{sales centers} \tag{6.6}$$

$$\sum_{i \in P, a \in I: i \neq a} Z_{iak} - \sum_{a \in I, j \in P: a \neq j} Z_{ajk} = 0, \forall k \in K \tag{6.7}$$

$$Z_{ijk} \leq M_k, \forall i, j \in I, k \in K \tag{6.8}$$

$$\sum_{m \in P: i \neq m} Z_{imk} + \sum_{n \in P: n \neq j} Z_{jnk} - Y_{ij} = 1, \forall i \in I, j \in J, k \in K \tag{6.9}$$

$$\sum_{j \in P: i \neq j} U_{ji} - \sum_{j \in P: i \neq j} U_{ij} = dd_i, \forall i \in I \tag{6.10}$$

$$U_{ij} \leq q_k Z_{ijk}, \forall i, j \in P, k \in K : i \neq j \tag{6.11}$$

$$YY_{jk} \leq YY_{ik} - cc\left(\frac{em_k}{q_k}\right) \times d2_{ij} Z_{ijk} + tt_k\left(1 - Z_{ijk}\right), \forall i, j \in P, k \in K : i \neq j \tag{6.12}$$

$$YY_{jk} \leq tt_k - cc\left(\frac{em_k}{q_k}\right) \times d2_{ij} Z_{ijk}, \forall i, j \in P, k \in K : i \neq j \tag{6.13}$$

Minimizing the costs involved in serving customers starting from the suppliers is given in Eq. (6.1) by a group of three summations, where the first summation minimizes the distance between the supplier and the sales center. The second block optimizes a fixed cost, $j \in J$, to ensure the respective sales center is operational. Finally, the third block of summation, considering all candidate routes that can be built, finds the best route that will have the most appropriate cost, i.e., a logistic network is made feasible. Objective (6.2) quantifies the fuel used by the distributor vehicle and the amount of CO_2 generated during the trip to deliver products to customers.

Condition (6.3) models the supply process from the supplier to the sales center, provided that the capacity of the sales center is not exceeded and that it is open to receive products since it has been located at a previous stage. In (6.4), the supplier supplies products to the sales center, following estimates of demands generated in the consumer market due to customer participation. The constraints (6.5) and (6.6) model customer service through a vehicle that reaches a customer i and then continues with another customer j.

The constraint (6.7), called the equilibrium condition, guarantees that as soon as a customer is served, the vehicle continues its journey to deliver a product to another customer and so on until a service is provided to all customers, following the routing plan built by the model. In (6.8), the customers are guaranteed to be served through a routing, as long as a vehicle is available for that purpose. Condition (6.9) guarantees that following a routing process, the sales center will serve the demand of customers using a vehicle. In (6.10) the delivery of products to a customer happens before serving the next customer, i.e., timely attention from the sales center to the customer. Constraint (6.11) ensures that routing to serve customers is possible, as long as the load in the vehicle, according to its capacity, has sufficient products.

Equation (6.12) models the fuel levels used in the vehicle's tank, when it arrives at the customer j to deliver the requested order. Finally, constraint (6.13) establishes that the fuel tank will be full if the vehicle has visited a fuel station and can therefore start delivering products to customers following an open vehicle routing.

6.4.2 Hybrid Algorithm

Different algorithmic strategies are used in the literature to propose acceptable solutions to the vehicle routing problem, such as exact methods, heuristics, meta-heuristics, machine learning clustering techniques, and hybrid algorithms, among others [1, 11, 17, 19, 38].

Based on the two objectives of this research, we developed a hybrid algorithm that combines the k-means algorithm and local search. Algorithms 1, 2, and 3 show the process of the methodology to be followed to perform vehicle routing. In the first part of Algorithm 1, the variables and objects are initialized reading a .txt file containing the data of customers, sales centers, and suppliers (lines 1–2), and the start time of the algorithm is calculated in line 3. Then, the number of clusters is set to 2 because the algorithm shows good performance when grouping the clusters and impacts positively on the vehicle routing (line 4). The k-means algorithm groups customers, sales centers, and suppliers into two clusters, minimizing the sum of distances between each one and the centroid of its group (lines 5–10). The clusters are then ordered in ascending order and the customers, sales centers, and suppliers (lines 11–13).

The first sales center in the list is chosen, the list of used clusters is updated according to the cluster number of the sales center, and the maximum value that will represent the current cluster is obtained; this also applies when kClusters >2 (lines 14–15). Then, the customers belonging to the current cluster are sorted according to their proximity distance to the current sales center (line 16). Likewise, the variables to be used in the creation of the arcs in the routing are initialized (lines 17–24).

The customer selection process is carried out according to their demand and the selection of the sales centers to satisfy this demand. Line 25 validates that all customers are selected according to their quantity, and the route object is initialized (line 26). There are two cases: when the initial node of the arc is a sales center and the final node is a customer (lines 27–37) and when the initial and final nodes of the arc are customers (lines 38–47). In both cases the initial node and its type (customer or sales center) and the list of used sales centers (when a new one is chosen) are updated, the next customer is validated and obtained from the validate_customer_selected method, and the QtyLoad is updated according to the demand of the next customer. The fuel variable of line 49 is assigned the result of the fuel consumption, the CO_2 emission according to the distance traveled, and the fuel leftover variable of line 3 of the second part of Algorithm 2 is updated.

There is a restriction to evaluate in the second part of Algorithm 2 in line 4, it is validated that if the vehicle capacity was exceeded or the fuel ran out, a new sales center, customer, and a new vehicle are chosen to make the trip, as well as the variables corresponding to the creation of the arc are updated, and it is validated that the selected customer is within the cluster of the sales center or in the next cluster (lines 5–29).

Finally, lines 30–48 correspond to the assignment of the values to each attribute of the route object, and then this route object is added to the route_N2 list. The loop ends when all customers have been included in the route.

The hybrid algorithm with level 2 was initiated in order to obtain the sales centers selected in the level 2 routing, as well as the amount of customer demand, which serve as inputs for Algorithm 3. This latter represents the routing process in level 1 between sales centers and suppliers. The algorithm starts with the initialization of the variables of the route_N1, and the variable unique_new_supplier_list_order returns the ascending ordered list of unique supplier clusters (lines 1–2). Then, a loop is performed to validate that for all sales centers the respective supplier has been found to satisfy their demands (line 3). For this, the cluster number of the sales center is obtained, as well as its x and y coordinates in the mapping plane (lines 4–6). The route object is initialized (line 7), and the while loop validates if there are suppliers in the list (line 8).

First, the closest unused supplier to the current sales center cluster is searched according to its distance between them and stored in the distance_f_p list (lines 9–16). If the distance_f_p list is not empty, it is sorted in ascending order, and the first supplier is selected, ending the loop; otherwise, the next cluster is extracted from the list supplier from the unique_new_supplier_list_order and updated (lines 17–26). At the end of the loop, the data of the attributes of the route object are updated, and the arc is saved in the route_N1 list to be returned by the algorithm (lines 27–38 and 40). Finally, the total time to execute the level 1 and 2 routes is calculated (line 39).

Algorithm 1 Part 1 of the hybrid algorithm—level 2

```
Require: Collection points of customers, sales centers, suppliers, and homogeneous fleet of vehicles.
Ensure: Minimum number of vehicles, minimum distance traveled to satisfy the customer and sales centers demand and minimum co2 emission.
Initialize the objects of customers, sales_centers, suppliers, and trucks when reading the FileTxt.
Initialize variables kClusters, customersCluster, salesCenterCluster, supplierCluster, truckId, Totalfuel, QtyDistancy, QtyCO2, QtyLoad, routes, routeIndex, route_N2
t = time.time()
kClusters = 2    ▷ Initialize the clusters using the kmeans algorithm
kmeans = KMeans(n_clusters=kClusters, random_state=0).fit(np.concatenate((salesCenter_coords, customers_coords, suppplier_coords)))
arr_cluster = kmeans.cluster_centers_
arr_sorted = arr_cluster[arr_cluster[:, 0].argsort()]    ▷ Sort the clusters in ascending order
customersCluster = pairwise_distances_argmin_min(customers_coords, arr_sorted)[0]
salesCenterCluster = pairwise_distances_argmin_min(salesCenter_coords, arr_sorted)[0]
supplierCluster = pairwise_distances_argmin_min(suppplier_coords, arr_sorted)[0] ▷ Update the cluster_number in the objects designated as customers, Sales_centers and Suppliers
salesCenterList = sort_salesCenter_cluster()
customersList = sort_customers_cluster()    ▷ sort the customers based on the cluster number
supplierList = sort_supplier_cluster()    ▷ sort the suppliers based on the cluster number
salesCenterSelect = SalesCenter[salesCenterList[idx_facil][0]]    ▷ select the first sales center of the list
num_clust = validate_cluster(list_clusters, salesCenterSelect.get_num_clust())    ▷ update the list of used clusters and get the maximum number of clusters
dist_min = sort_customers(customers, salesCenterSelect, num_clust)    ▷ sort customers close to the actual sales center
routes_N2 = []    ▷ start generating the routes from sales centers to customers
truck_id = 0
QtyLoad = 0
Qtydist = 0
productToDeliver = trucks[truck_id].get_capacity()
QtyCO2 = 0
fuel_total = trucks[truck_id].get_fuel()
ruta_inx = 0
while qty_customer_used < num_customers do    ▷ validate if all customers are used in the routes
    route = Route()
    if route_inx == 0 then    ▷ if the arc is formed between sales center and customer
        route.set_type_ini("sales center")
        node_ini = salesCenter_select.get_index()
        list_salesCenterUsed.append(salesCenter_select.get_index())
        band, index, pos_customer, dist, qty_customer_used, dist_min, list_clusters = validate_customer_selected(list_clusters, salesCenter, dist_min, qty_customer_used, route_inx) ▷ Validates if the selected customer was taken, also if all customers in the cluster have been selected, choose from the next cluster.
        if band == True then
            route_inx = index
        end if
        node_end = customer[pos_customer].get_index()
        route.set_type_dest("customer")
        QtyLoad += customer[pos_customere].get_demand()
    else    ▷ if the arc is formed between customers
```

```
        route.set_type_ini("customer")
        node_ini = customer[pos_customer].get_index()
        band, index, pos_customer, dist, qty_customer_used, dist_min, list_clusters = validate_customer_selected(list_clusters, customers, dist_min, qty_customer_used, route_inx)
    end if
    fuel = (trucks[truck_id].get_fuel_consumption() *(trucks[truck_id].get_emission_co2()/trucks[truck_id].get_fuel()) * dist)
Continue processing
end while
```

Algorithm 2 Part 2 of the hybrid algorithm—level 2

Require: Collection points of customers, sales centers, suppliers, and homogeneous fleet of vehicles.

Ensure: Minimum number of vehicles, minimum distance traveled to satisfy the customers and sales centers demand and minimum co2 emission.

```
while qty_customers_used < num_customers do    ▷ validate if all customers are used in the routes
Continue processing
    fuel_leftover = fuel_total - fuel
    if QtyLoad > trucks[truck_id].get_capacity() or fuel_total <= 0 or fuel_leftover <= 0 then    ▷ if you run out of fuel, you choose another vehicle and a new facility to tour the remaining customers
        customers[pos_customer].set_used(0)
        productToDeliver = trucks[truck_id].get_capacity()
        qty_customer_used -= 1    ▷ one is removed because a new target customer is chosen
        truck_id += 1
        QtyCO2 = 0
        fuel_total = trucks[truck_id].get_fuel()
        route.set_vehicle(truck_id)
        route_inx = 0
        idx_facil += 1    ▷ we choose a new sales center
        salesCenter_select = salesCenter[list_salesCenter_order[idx_facil][0]]
        node_ini = salesCenter_select.get_index()
        route.set_type_ini("sales center")
        list_salesCenter_used.append(salesCenter_select.get_index())
        num_clust = validate_cluster(list_clusters, salesCenterSelect.get_num_clust())
        dist_min = sort_customers(customers, salesCenterSelect, num_clust)
        band, index, pos_customer, dist, qty_customer_used, dist_min, list_clusters = validate_customer_selected(list_clusters, customers, dist_min, qty_customer_used, route_inx)
        if band == True then
            route_inx = index
        end if
        node_end = customer[pos_customer].get_index()
        route.set_type_dest("customer")
```

```
            QtyLoad += customer[pos_customer].get_demand()
            fuel = (trucks[truck_id].get_fuel_consumption()
    *(trucks[truck_id].get_emission_co2()/trucks[truck_id].get_fuel()) * dist)
            fuel_leftover = fuel_total - fuel
        end if
        Qty_demand_customers += customers[pos_customer].get_demand()
        route.set_productToDeliver(productToDeliver)
        productToDeliver -= customers[pos_customer].get_demand()
        route.set_vehicle(truck_id)
        route.set_index(route_inx)
        route.set_customer_ini(node_ini)
        route.set_customer_dest(node_end)
        route.set_load(QtyLoad)
        cost_flete = trucks[truck_id].get_flete() * dist / 100
        route.set_cost_flete_salesCenter_customer(cost_flete)
        Qtydist += dist
        route.set_route(dist)
        c_co2 = trucks[truck_id].get_fuel_consumption() *
    (trucks[truck_id].get_emission_co2()/trucks[truck_id].get_fuel() * dist)
        QtyCO2 += c_co2
        route.set_co2(QtyCO2)
        route.set_fuel(fuel_total )
        fuel_total = fuel_total - fuel
        route_N2.append(route)
        route_inx += 1
    end while
    elapsed = time.time() - t                              ▷ Total algorithm time.
    return route_N2
```

6.5 Computational Results and Discussion

6.5.1 *Results Using the Optimization Model*

The potential of the proposed model was demonstrated when it was implemented using the GLPK computational tool (https://www.gnu.org/software/glpk), which for the data instance is described in Table 6.2. In addition, a set of 4 suppliers, 4 sales centers, and 19 customers located in their respective geographical areas to receive the distributor vehicle. Each vehicle has 20 liters of fuel when it starts its trip. Finally, the CO_2 emission varies as the fuel is used according to the distance traveled.

Table 6.3 shows the supply of products to the sales centers $f1$ and $f2$ by the suppliers located $s1$ and $s3$ and additionally to the routes from $s1-f1$ and $s3-f2$ which were 25 and 17.49 km. The demand of the sales centers, which are 36 units of products, is in direct relation with the demand of the customers; however, for the supply to the customers, this quantity was divided and assigned to the sales centers at level 2 which will attend the customers following an open vehicle routing process.

Algorithm 3 Hybrid algorithm—level 1

```
Require: Collection points of sales centers and suppliers, supplier_list_order, salesCenter_select, Qty_demand_customers.
Ensure: Suppliers that satisfy sales centers demand
Initialize route_N1
unique_new_supplier_list_order = unique_supplier_list_order()
for id_facil = 0: id_facil < len(salesCenter_select) : id_facil ++ do
    id_salesCenter_cluster = salesCenter[salesCenter_select[id_facil]].get_cluster_number()
    x1 = salesCenter[salesCenter_select[id_facil]].get_x_coord()
    y1 = salesCenter[salesCenter_select[id_facil]].get_y_coord()
    route = Route()
    while len(unique_new_supplier_list_order) ! = 0 do
        cluster_supplier= find_nearest(unique_new_supplier_list_order, id_salesCenter_cluster)
        distance_f_p = []
        for id_supplier = 0: id_supplier < len(supplier_list_order) : id_supplier ++ do
            if supplier_list_order[id_supplier][1] == cluster_supplier and supplier[supplier_list_order[id_supplier][0]].get_used() == 0 then
                x2 = supplier[supplier_list_order[id_supplier][0]].get_x_coord()
                y2 = supplier[supplier_list_order[id_supplier]0]].get_y_coord() distance_f_p.append([distance(x1, y1, x2, y2), supplier[supplier_list_order[id_supplier][0]].get_index()])
            end if
        end for
        if len(distance_f_p) > 0 then
            distance_f_p.sort()
            dist_F_P= distance_f_p[0][0]
            supplier_select = distance_f_p[0][1]
            supplier[supplier_select].set_used(1)
            break
        else
unique_new_supplier_list_order = np.setdiff1d(unique_new_supplier_list_order, np.array([cluster_supplier])) n = len(unique_new_supplier_list_order)
        end if
    end while
    route.set_type_ini("supplier")
    route.set_customer_ini(supplier_select)
    route.set_vehicle(truck_id)
    route.set_type_dest("sales center")
    route.set_customer_dest(salesCenter[salesCenter_select[id_facil]].get_index())
    route.set_load(Qty_demand_customers)
    route.set_route(distance_f_p[0][0])
    route.set_indice(route_ind)
    route_ind += 1
    truck_id += 1
    route_N1.append(route)
end for
elapsed = time.time() - t                                  ▷ Total algorithm time.
return route_N1
```

Table 6.2 Location of suppliers, sales centers, and customers

Entity	Supplier	Sales center: capacity	Customers: demand
s_1	(35, 15)		
s_2	(100, 65)		
s_3	(7, 59)		
s_4	(5, 65)		
f_1		(10, 15) : 120	
f_2		(22, 50) : 120	
f_3		(30, 70) : 120	
f_4		(55, 105) : 120	
c_1			(5, 35) : 4
c_2			(20, 61) : 3
c_3			(28, 80) : 3
c_4			(40, 90) : 1
c_5			(80, 50) : 3
c_6			(18, 30) : 1
c_7			(35, 25) : 5
c_8			(75, 40) : 2
c_9			(30, 40) : 1
c_{10}			(18, 40) : 1
c_{11}			(10, 50) : 2
c_{12}			(45, 27) : 1
c_{13}			(40, 32) : 2
c_{14}			(48, 19) : 1
c_{15}			(5, 30) : 2
c_{16}			(40, 52) : 1
c_{17}			(45, 38) : 1
c_{18}			(23, 70) : 1
c_{19}			(28, 85) : 1

Table 6.3 Computational result at level 1: supplier-sales center

Supplier: sales center	Sales centers demand	Travel (km)
$s1-f1$	36	25
$s3-f2$	36	17.49

At level 2, after 4.28 minutes of computational process used for the implementation of the proposed model, it was able to obtain the route plan to be followed to supply products to the customers as shown in Table 6.4, for which it was necessary to select two vehicles that transport the orders to the customers through open vehicle routing. In addition, the cost of serving the customer in monetary units is also obtained. Finally, it is determined how the fuel, in liters, in the tank of the distributor vehicles was spent during the trip and consequently to quantify the amount of CO_2 generated in g/L.

Table 6.4 Computational result at level 2: sales center-customers

Tour	Product to be delivered	Distance (km)	Cost	Tank (l)	CO_2 (g/L)
$f1-c6$	18	17	8.5	20	0.89
$c6-c15$	17	13	6.5	19.1	0.68
$c15-c1$	15	5	2.5	18.42	0.26
$c1-c11$	11	15.81	7.9	18.16	0.83
$c11-c2$	9	14.86	7.43	17.33	0.78
$c2-c18$	6	9.48	4.74	16.55	0.49
$c18-c3$	5	11.18	5.59	16.05	0.58
$c3-c19$	2	5	2.5	15.46	0.26
$c19-c4$	1	13	6.5	15.2	0.68
$f3-c10$	18	10.77	5.38	20	0.56
$c10-c9$	17	12	6	19.43	0.63
$c9-c16$	16	15.62	7.81	18.8	0.82
$c16-c17$	15	14.86	7.43	17.98	0.78
$c17-c13$	14	7.81	3.9	17.2	0.41
$c13-c7$	12	8.6	4.3	16.79	0.45
$c7-c12$	7	10.19	5.09	16.34	0.53
$c12-c14$	6	8.54	4.27	15.8	0.44
$c14-c8$	5	34.2	17.1	15.35	1.79
$c8-c5$	3	11.18	5.59	13.56	0.58

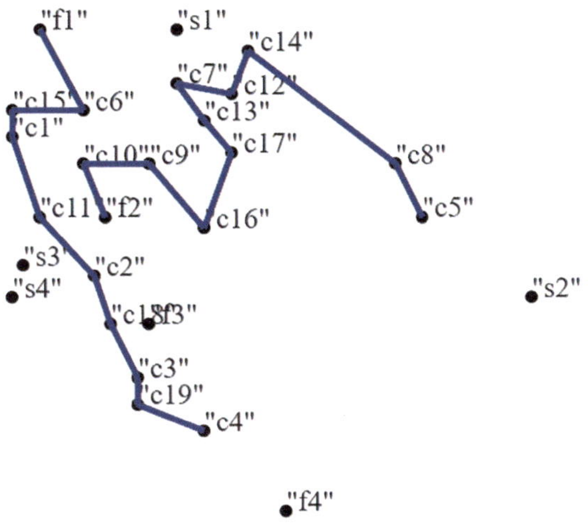

Fig. 6.5 Location-routing of open vehicles

For a more clear visual representation, Table 6.4 description is also shown in Fig. 6.5, where two paths can be seen starting at the sales centers $f1$ and $f2$ and ending at the customers $c4$ and $c5$. Finally, the cars will be empty and depart to their depots.

Figure 6.6 shows the objective functions for cost and fuel consumption in the proposed model. In each interval of the constructed route, the costs vary in value

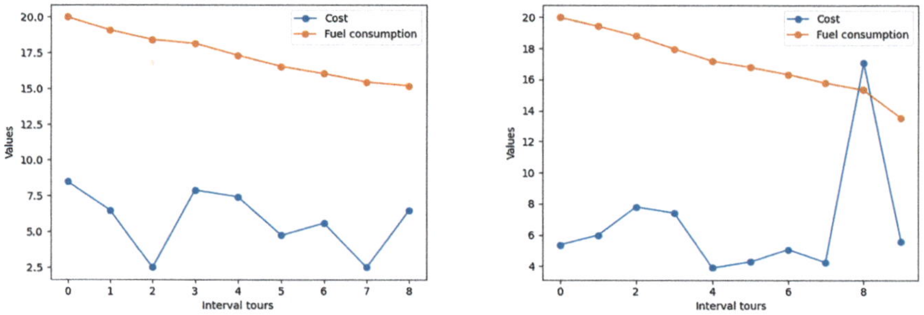

Fig. 6.6 Comparison of the objective functions for the optimization model. (**a**) Level 2: Route 1. (**b**) Level 2: Route 2

according to the distance traveled, which implies fuel consumption. Therefore, for each of the two routes, the cost is economically minimized to find the best route. This means saving of the fuel used to deliver the products and consequently the amount of CO_2 generated, thus contributing to the environment of the urban area where it is applied.

6.5.2 Results Using the Hybrid Algorithm

The implementation of the proposed hybrid algorithm was alongside in Python programming language in Google Colaboratory alongside the sklearn, numpy, and matplotlib libraries. Also, three test cases were used to evaluate the performance of the hybrid algorithm, as presented below.

6.5.2.1 Case 1

Based on the information in Table 6.2, the proposed hybrid algorithm was tested for its effectiveness. After the implementation was completed for this scenario, the algorithmic procedure quickly generated results as shown in Tables 6.5 and 6.6 and Fig. 6.7.

Table 6.5 presents two transportation routes: $s1-f1$ and $s3-f2$, whose routes were 25 km and 15.81 km, respectively, and Table 6.6 details the routes between sales centers and customers. In this case, the vehicles started the trip with 20 liters of fuel and a capacity of 18 weights units. At the second level, two vehicles were required to meet the demand of customers, and their distance traveled was 382.42 km. Finally, the total distance traveled across both levels was 423.24 km, and the execution time of the hybrid algorithm to find results was 0.014 seconds.

Table 6.5 Result at level 1 with hybrid algorithm: supplier-sales center

Supplier: sales center	Demand of sales center	Route (km)
$s1-f1$	36	25.00
$s3-f2$	36	15.81

Table 6.6 Result at level 2 with hybrid algorithm: sales center-customers

Tour	Product to be delivered	Distance (km)	Cost	Tank (l)	CO_2 (g/L)
$f1-c15$	18	15.81	3.95	20.00	0.74
$c15-c6$	16	13.00	3.25	19.25	0.61
$c6-c1$	15	13.93	3.48	18.64	0.65
$c1-c10$	11	13.93	3.48	17.98	0.65
$c10-c7$	10	22.67	5.67	17.32	1.07
$c7-c9$	5	15.81	3.95	16.25	0.74
$c9-c13$	4	12.80	3.20	15.50	0.61
$c13-c11$	2	34.99	8.75	14.90	1.65
$f2-c2$	18	11.00	2.75	20	0.52
$c2-c16$	15	21.93	5.48	19.48	1.04
$c16-c18$	14	24.76	6.19	18.44	1.17
$c18-c17$	13	38.83	9.71	17.27	1.83
$c17-c12$	12	11.00	2.75	15.44	0.52
$c12-c14$	11	8.54	2.14	14.92	0.40
$c14-c8$	10	34.21	8.56	14.52	1.61
$c8-c5$	8	11.18	2.80	12.90	0.53
$c5-c3$	5	60.03	15.01	12.37	2.84
$c3-c19$	2	5.00	1.25	9.53	0.24
$c19-c4$	1	13.00	3.25	9.30	0.62

Figure 6.7 shows two clusters represented by different colors, which served as the input for vehicle routing. First, the sales center was selected, after the customers belonging to the first cluster were then selected following the constraints of the problem. Likewise, the second vehicle's route included some customers left over from the first cluster along with the customers from the second cluster. Finally, suppliers were selected to meet the demand of the selected sales centers, which required 36 units of products, corresponding to the customer demand satisfied in each route made by each vehicle.

Figure 6.8 presents a comparison between the objective functions used to determine the cost of developing the two routes and the corresponding fuel to use, revealing a decrease in the amount of fuel in the delivery vehicle's tank as the route plan progresses. Therefore, since the objective is to reduce costs to serve customers, this implies a reasonable use of the fuel that will be used, and consequently, a reduction in the amount of CO_2 generated will occur, benefiting the urban area where it is applied.

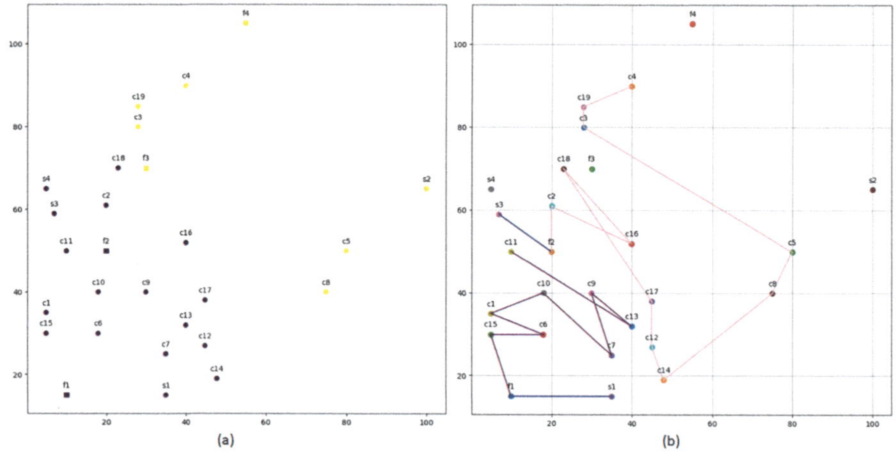

Fig. 6.7 (**a**) Clusters formed with Table 6.2 data using k-means algorithm. (**b**) Open vehicle routing applying the hybrid algorithm

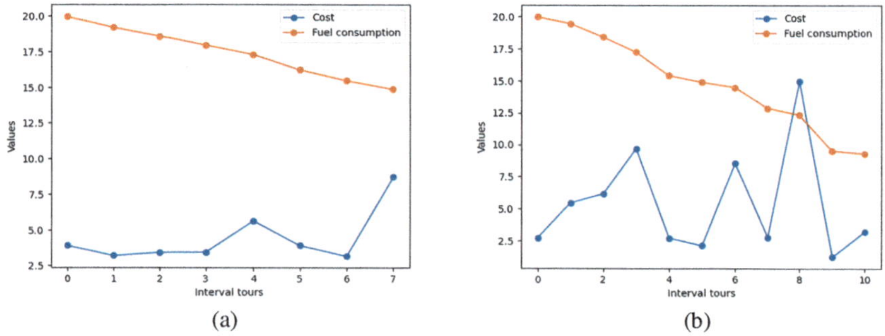

Fig. 6.8 Comparison of the objective functions for the hybrid algorithm. (**a**) Level 2: Route 1. (**b**) Level 2: Route 2

6.5.2.2 Case 2

In this case, a new scenario was formed with 30 customers, 10 sales centers, and 10 suppliers. When the hybrid algorithm was executed, two clusters were formed based on the coordinates of the components in case 2, as well as the routes of the vehicles as shown in Fig. 6.9. The total distance traveled in both levels was 605.21 km, and the execution time of the hybrid algorithm was 0.018 seconds.

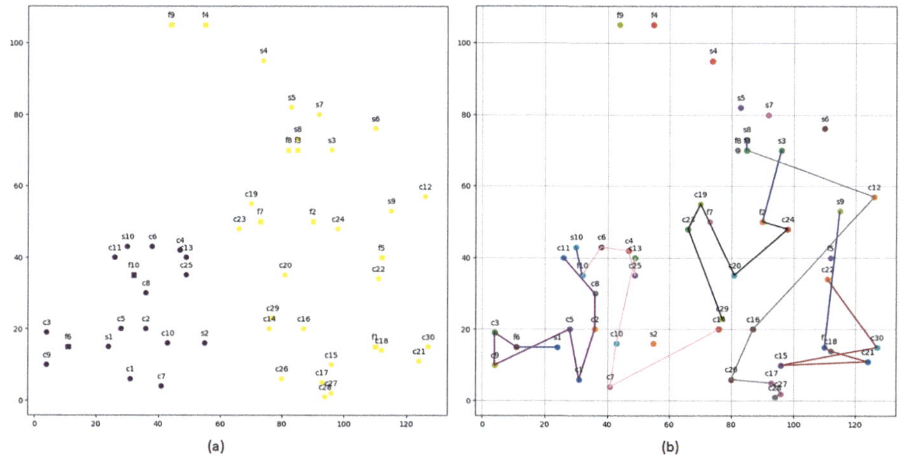

Fig. 6.9 (**a**) Clusters formed for case 2 through the k-means algorithm. (**b**) Result of open vehicle routing applying the Hybrid algorithm

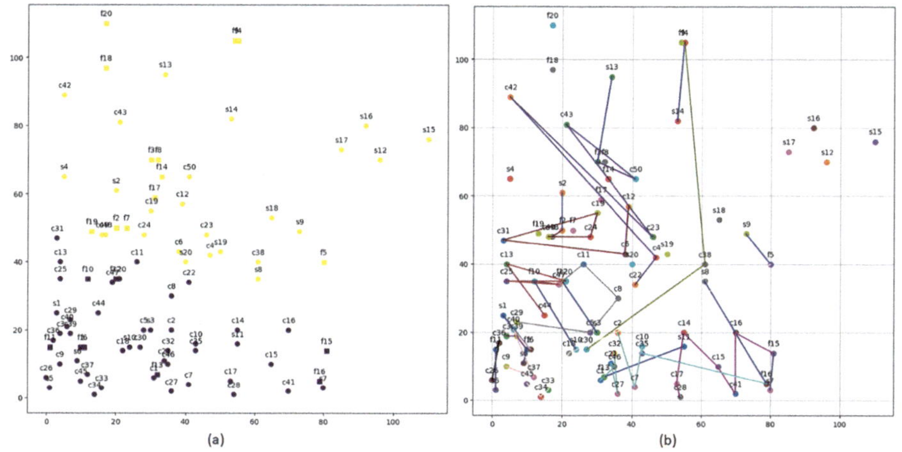

Fig. 6.10 (**a**) Clusters formed in case 3 using the k-means algorithm. (**b**) Result of open vehicle routing applying the hybrid algorithm

6.5.2.3 Case 3

A third scenario consisting of 50 customers, 20 sales centers, and 20 suppliers was constructed. With this information, two clusters and their respective routes were obtained as shown in Fig. 6.10, with a total distance of 1070.84 km. To obtain the results, the execution time of the hybrid algorithm was 0.020 seconds.

6.6 Perspectives on the Circular Economy in the Urban Context

The proposal discussed in Sect. 6.4 is interesting and important as its proper use determines an organized consumption of various products on a massive scale, which after use generates waste that must be optimally collected for subsequent recycling or reuse or disposal. The first two types, which are considered to have value, are transported to specialized centers, while the third to sanitary landfills.

The collection of valuable waste is a challenge for municipal authorities, since the collected waste has to return to the productive cycle in some form for later use and consequently activates the circular economy contributing on the one hand to the non-excessive use of natural resources. According to the United Nations, the "Demand for natural resources is at an all-time high and continues to grow - for food, clothing, water, housing, infrastructure and other aspects of life" (https://www.un.org/en/actnow/facts-and-figures) has a negative impact on the planet since the excessive processes of extracting and processing materials, fuels, and food are responsible for half of the total global greenhouse gas emissions and more than 90% of the loss of biodiversity and water stress.

On the other hand, this type of economy generates new sources of employment and consequently an increase in the purchasing power of the working masses for their family welfare, since it improves their living status by being able to buy or sell according to the urban activity they carry out. In addition, the circular economy contributes to the protection of the environment through the proper use of reverse logistics as it is a more resilient and sustainable model that helps to reduce greenhouse gas emissions and innovation in waste management as already discussed.

Contributions of the circular economy are presented in various articles such as solid waste management to achieve the Sustainable Development Goals (SDGs) discussed in [31]. An investigation of circular economy-based factors contributing to the sustainable development of smart cities is presented in [8]. More recently in [2] they present the impact on supply chain and circular economy initiatives in small enterprises, based on the implementation of sustainable circular supply chain activities and their influence on circular economic, social, and environmental factors. Therefore, the traditional economic model (linear economy) based on taking, making, buying, consuming, and discarding products that damage natural resources and impact negatively on the environment and then gradually replaced by the circular economy concept is a great contribution to maintaining the sustainability of the planet and, consequently, a great contribution for future generations who should receive a more sustainable environment healthy with a minimum of contaminations.

6.7 Conclusions

The chapter shows that it is possible to contribute to the supply of goods in urban areas by developing a bi-objective optimization model and a hybrid algorithm. The routing activity is performed for the delivery of goods to customers. In addition, to contribute to the environment, the proposal models fuel consumption and measures the CO_2 emission that occurs in the customer service process in the urban area where it is applied.

The proposal is based on the premise that the locations of an estimated number of customers in a geographic area are known, and when these customers generate a demand for the sales centers, a supply is produced that starts at the suppliers, as shown in the results. This premise contributes to reducing service and transportation costs in the selecting optimal routes, with the consequent decrease in fuel use and therefore in CO_2 emission.

The results presented show the potential of the proposed model, as solutions for scenarios with 19 or fewer customers are obtained quickly. However, when considering a scenario of 20 customers, the processing time exceeded the established limit of 60 minutes, this is due to the fact that routing is an NP-hard problem. The limitation of the proposed model was addressed by incorporating k-means clustering for scenarios of 30 and 50 customers, with which good solutions are obtained in reasonable times.

References

1. Abdellaoui, A., Benabbou, L., Hallaoui, I. E.: Towards a connection between the capacitated vehicle routing problem and the constrained centroid-based clustering. arXiv preprint arXiv:2403.14013 (2024). https://doi.org/10.48550/arXiv.2403.14013
2. Alamelu, R., Jayanthi, M., Dinesh, S., Nalini, R., Shobhana, N., Amudha, R: Sustainable supply chain and circular economy ingenuities in small manufacturing firms- a stimulus for sustainable development. Material today: Proceedings, Science Direct, Elsevier **92**, 17–23 (2023). https://doi.org/10.1016/j.matpr.2023.03.236
3. Amiri, A., Amin, S. H., Zolfagharinia, H.: A bi-objective green vehicle routing problem with a mixed fleet of conventional and electric trucks: Considering charging power and density of stations. Expert Systems with Applications, Science Direct, Elsevier **213**, Part C (2023). https://doi.org/10.1016/j.eswa.2022.119228
4. Bouchery, Y., Corbett, C. J., Fransoo, J.C., Tan, T.: Sustainable supply chains: A research - based textbook on operations and strategy. Springer International publishing, Switzerland (2017). https://doi.org/10.1007/978-3-031-45565-0
5. Brandao, J.: Iterated local search algorithm with ejection chains for the open vehicle routing problem with time windows. Computers & Industrial Engineering, Science Direct, Elsevier **120**, 146–159 (2018). https://doi.org/10.1016/j.cie.2018.04.032
6. Cao, J. X., Zhang, Z., Zhou, Y.: A location-routing problem for biomass supply chains. Computers & Industrial Engineering, Science Direct, Elsevier **152**(107017) (2021). https://doi.org/10.1016/j.cie.2020.107017
7. Chen, W., Zhang, D., Van Woensel, T., Xu, G., Guo, J.: Green vehicle routing using mixed fleets for cold chain distribution. Experts Systems with Applications, Science Direct, Elsevier **233**(120979) (2023). https://doi.org/10.1016/j.eswa.2023.120979

8. Dincă, G., Milan, A-A., Andronic, M. L., Pasztori, A-M, Dină, D.: Does Circular Economy contribute to smart cities' sustainable development?. International Journal of Environmental Research and Public Health, MDPI **19**(13), 7627 (2022). https://doi.org/10.3390/ijerph19137627
9. Donthu, N., Kumar, S., Mukherjee, D., Pandey, N., and Marc Lim, W.: How to conduct a bibliometric analysis: An overview and guidelines. Journal of Business Research, Science Direct, Elsevier **133**, 285–296 (2021). https://doi.org/10.1016/j.jbusres.2021.04.070
10. Elatar, S., Abouelmehdi, K., Riffi, M. E.: The vehicle routing problem in the last decade: variants, taxonomy and metaheuristics. The 14th International Conference on Ambient Systems, Networks and Technologies (ANT) March 15–17, 2023, Leuven, Belgium. Procedia Computer Science, Science Direct, Elsevier **220**(2023), 398–404 (2023). https://doi.org/10.1016/j.procs.2023.03.051
11. Frías, N., Johnson, F., Valle, C.: Hybrid Algorithms for energy minimizing vehicle routing problem: integrating clusterization and ant colony optimization. IEEE Access. (2023) https://doi.org/10.1109/ACCESS.2023.3325787
12. Ghorbani, E., Alinaghian, M., Gharehpetian, G. B., Mohammadi, S., Perboli, G.: A Survey on Environmentally Friendly Vehicle Routing Problem and a Proposal of Its Classification. Sustainability, MDPI **12**(21), 9079 (2020). https://doi.org/10.3390/su12219079
13. Govindan, K., Nosrati-Abarghooee, S., Nasiri, M. M., Jolai, F.: Green reverse logistics network design for medical waste management: A circular economy transition through case approach. Journal of Environmental management, Science Direct, Elsevier **322**(2022), 115888 (2022). https://doi.org/10.1016/j.jenvman.2022.115888
14. Govindan, K., Salehian, F., Kian, H., Teimoor Hosseini, S.: A location-inventory-routing problem to design a circular closed-loop supply chain network with carbon tax policy for achieving circular economy: An augmented epsilon-constraint approach. International Journal of Production Economics, Science Direct, Elsevier **257**(2023), 108771. https://doi.org/10.1016/j.ijpe.2023.108771
15. Heidari, A., Imani, D. M., Khalilzadeh, M., Sarbazvatan, M.: Green two-echelon closed and open location-routing problem: application of NSGA-II and MOGWO metaheuristic approaches. Environment, Development and Sustainability, Springer **25**, 9163–9199, (2023). https://doi.org/10.1007/s10668-022-02429-w
16. Hajghani, M., Forghani, M. A., Heidari, A., Khalilzadeh, M., Kebriyaii, O.: A two-echelon location routing problem considering sustainability and hybrid open and closed routes under uncertainty. Heliyon, ScienceDirect **9**(3), e14258 (2023). https://doi.org/10.1016/j.heliyon.2023.e14258
17. Han, X.: Path planning algorithm for the multiple depot vehicle routing problem based on parallel clustering. Scientific Programming, 2023(1), 7588595 (2023). https://doi.org/10.1155/2023/7588595
18. Idzikowski, R.: Capacitated Open Vehicle Routing Problem with Time Couplings. In: Zamojski, W., Mazurkiewicz, J., Sugier, J., Walkowiak, T., Kacprzyk, J. (eds), Theory and Applications of Dependable Computer Systems. DepCoS-RELCOMEX 2020. Advances in Intelligent Systems and Computing, vol 1173, pp. 273–282. Springer, Cham (2020). https://doi.org/10.1007/978-3-030-48256-5_27
19. Jiao, L., Peng, Z., Xi, L., Guo, M., Ding, S., Wei, Y.: A multi-stage heuristic algorithm based on task grouping for vehicle routing problem with energy constraint in disasters. Expert Systems with Applications, 212, 118740 (2023). https://doi.org/10.1016/j.eswa.2022.118740
20. Lyu, P., Wang, P.S., Liu, Y., Wang, Y.: Review of the studies on emission evaluation approaches for operating vehicles. Journal of traffic and transportation engineering (English edition), Science Direct, Elsevier **8**(4), 493–509 (2021). https://doi.org/10.1016/j.jtte.2021.07.004
21. Kumar, M., Kumar, D., Saini, P., Pratap, S.: Inventory routing model for perishable products toward circular economy. Computer & Industrial Engineering, Science Direct, Elsevier **169**, 108220 (2022). https://doi.org/10.1016/j.cie.2022.108220

22. Navazi, F., Sedaghat, A., Tavakkoli-Moghaddam, R.: A New Sustainable Location-Routing Problem with Simultaneous Pickup and Delivery by Two-Compartment Vehicles for a Perishable Product Considering Circular Economy. IFAC-PapersOnLine, Science Direct, Elsevier **52**(13), 790–795 (2019). https://doi.org/10.1016/j.ifacol.2019.11.212
23. Mallick, P. K., Salling, K. B., Pigosso, D. C. A., McAloone, T. C.: Closing the loop: Establishing reverse logistics for a circular economy, a systematic review. Journal of Environmental Management, Science Direct, Elsevier **328**, 117017 (2023). https://doi.org/10.1016/j.jenvman.2022.117017
24. Mansoure, B., Sahu, S., Ülkü, M. A.: Toward Greening City Logistics: A Systematic Review on Corporate Governance and Social Responsibility in Managing Urban Distribution Centers. Logistics, MDPI **7**(19), (2023). https://doi.org/10.3390/logistics7010019
25. Pedram, A., Sorooshian, S., Mulubrhan, F., Abbaspour, A.: Incorporating Vehicle-Routing Problems into a Closed-Loop Supply Chain Network Using a Mixed-Integer Linear-Programming Model. Sustainability, MDPI **15**(4), 2967 (2023). https://doi.org/10.3390/su15042967
26. Qie, B., Weng, X., Sun, Z., Jin, M., Yu, R.: Location-routing and cost-sharing models under joint distribution. Cluster Computing, Springer **27**, 5879–5891 (2024). https://doi.org/10.1007/s10586-024-04282-0
27. Rodriguez-Melquiades, J., Lujan, E., Segura, F.G.: An Integrated Model for Locating-Routing in the Goods Delivery and Simultaneous Pickup in the Urban Context. In: Osvaldo Gervasi et al. (Eds.), Computational Science and Its Applications – ICCSA 2021, Lecture Notes in Computer Science, vol. 12952, pp. 323–336. Springer, Heidelberg (2021). https://doi.org/10.1007/978-3-030-86973-1_23
28. Rodriguez-Melquiades, J., Ramirez Cerna, L.: Reverse logistics based on green open vehicle routing for waste collection in the urban context. Pesquisa operacional, Brazilian Operations Research Society **43**, (2023). https://doi.org/10.1590/0101-7438.2023.043.00269174
29. Ruiz, E., Soto, V., Ruiz, A. E., Reyes, R.: Solving the open vehicle routing problem with capacity and distance constraints with a biased random key genetic algorithm. Computers & Industrial Engineering, Science Direct, Elsevier **133**, 207–219 (2023). https://doi.org/10.1016/j.cie.2019.05.002
30. Schneider, M., Drexl, M.: A survey of the standard location-routing problem. Annals of Operations Research, Springer **259**, 389–414 (2017). https://doi.org/10.1007/s10479-017-2509-0
31. Sharma, H. B., Vanapalli, K. R., Samal, B., Cheela, V.R. S., Dubey, B. K., Bhattacharya, J.: Circular economy approach in solid waste management system to achieve UN-SDGs: Solutions for post-COVID recovery. Science of the Total Environment, Science Direct, Elsevier **800**, 149605 (2021). https://doi.org/10.1016/j.scitotenv.2021.149605
32. Shi, Y., Lin, Y., Wang, S., Wen, H., Lim, M. K., Li, Y.: A simultaneous facility location and vehicle routing problem with recyclable express packaging consideration for sustainable city logistics. Sustainable Cities and Society, Science Direct, Elsevier **98**, 104857 (2023). https://doi.org/10.1016/j.scs.2023.104857
33. Sluijk, N., Florio, A.M., Kinable, J., Dellaer, N., and Van Woensel, T.: Two-echelon vehicle routing problems: A literature review. European Journal of Operations Research, Science Direct, Elsevier **304**, 865–886 (2023). https://doi.org/10.1016/j.ejor.2022.02.022
34. Soto-Mendoza, V., Ruiz-y-Ruiz, E., García-Calvillo, I., Nucamendi-Guillen, S., Cardona-Valdés, Y.: A location-routing problem for local supply chains. Computers & Industrial Engineering, Science Direct, Elsevier **183**, 109528 (2023). https://doi.org/10.1016/j.cie.2023.109528
35. Stamadianos, T., Kyriakakis, N. A., Marinaki, M., Marinakis, Y.: The close-open mixed-fleet electric vehicle routing problem. Cleaner Logistics and Supply Chain, Science Direct, Elsevier **9**(2023), 100125 (2023). https://doi.org/10.1016/j.clscn.2023.100125
36. Stefanakis, A., Nikolaou, L.: Circular economy and sustainability, Volume 1: Management and policy. 1st Edition. Elsevier, September 14 (2021). https://doi.org/10.1016/C2019-0-00505-5

37. Stevels, A.: The challenge of introducing design for the circular economy in the electronics industry: A proposal for metrics. Circular Economy, Science Direct, Elsevier **2**(2023), 100051 (2023) . https://doi.org/10.1016/j.cec.2023.100051
38. Şehitoğlu, A., Aghayeva, C.: Comparison of Optimal Solutions of Clusters Created Using Clustering Algorithm with Meta-Heuristic Algorithms in Capacity Vehicle Routing Problem. In 2023 5th International Conference on Problems of Cybernetics and Informatics (PCI) (pp. 1–3). IEEE (2023).
39. Tadaros, M., Migdalas, A.: Bi- and multi-objective location routing problems: classification and literature review. Operational Research, Springer **22**, 4641–4683 (2022). https://doi.org/10.1007/s12351-022-00734-w
40. Tan, S-Y., Yeh, W-C.: The vehicle routing problem: State-of-the-Art Classification and Review. Applied Sciences, MDPI **11**(21), 10295 (2021). https://doi.org/10.3390/app112110295
41. Venes, H., Galavote, T., De Alvarenga Rosa, R., Siman, R. R.: "That's (not) Rubbish!" Planning the Reverse Logistic for Recyclable Solid Waste Using a Vehicle Routing Problem: A Case Study in Vitória/ES . International Journal of Sustainable Development and Planning, **18**(7), 2191–2198 (2023) . https://doi.org/10.18280/ijsdp.180723
42. Wang, Y., Luo, S., Fan, J., Zhen, L.: The multidepot vehicle routing problem with intelligent recycling prices and transportation resource sharing. Transportation Research Part E: Logistics and Transportation Review, Science Direct, Elsevier **185**, 103503 (2024) . https://doi.org/10.1016/j.tre.2024.103503
43. Xu, W., Zhang, C., Cheng, M., Huang, Y.: Electric Vehicle Routing Problem with Simultaneous Pickup and Delivery: Mathematical Modeling and Adaptive Large Neighborhood Search Heuristic Method. Energies, MDPI **15**(23), 9222 (2022). https://doi.org/10.3390/en15239222
44. Zhao, G.: Application of Swarm Intelligence Optimization Algorithm in Logistics Delivery Path Optimization under the Background of Big Data. Journal of Function Spaces, Hindawi **2023**, 347611 (2023). https://doi.org/10.1155/2023/3476711

Part II
Data Science for Smarter Circular Supply Chains

Chapter 7
Implementing Sustainability and Project Management Outcomes

Roya Anvari and Farideh Moradi

Abstract The integration of sustainability into project management is an evolving area within business management. This chapter conducts a comprehensive literature review on the intersection of sustainability and project management. Key topics explored include the impact on access to capital, customer loyalty, employee morale and productivity, supply chain improvements, long-term operational performance and efficiency, shareholder value enhancement, risk reduction, and talent recruitment and retention.

Keywords Sustainability · Project management · Business management · Operational performance risk management

7.1 Introduction

In the modern business environment, project success is defined not only by traditional metrics but also by the inclusion of sustainability principles. This chapter investigates the emerging field of sustainable project management and how it can enhance project outcomes by incorporating sustainability practices. The literature review addresses the benefits of sustainable project management on various business aspects, including capital access, customer loyalty, employee engagement, supply chain efficiency, operational performance, shareholder value, risk management, talent acquisition, and overall business resilience. Traditional project management aims to deliver projects within budget, on time, and within scope. However, in the twenty-first century, a holistic approach is essential, integrating sustainability into project management to consider environmental, social, and economic impacts

---

R. Anvari (✉)
School of Business and Administrative Studies, The University of Georgia, Tbilisi, Georgia
e-mail: r.anvari@ug.edu.ge

F. Moradi
Beykoz University, Istanbul, Turkey

B. Jarboui et al. (eds.), *Circular Economy Supply Chains*,
Computational Intelligence Methods and Applications,
https://doi.org/10.1007/978-981-96-7905-8_7

throughout the project life cycle [1]. This chapter reviews research on the positive impacts of sustainable project management practices.

7.2 Literature Review

Research increasingly supports the benefits of sustainable project management. Studies indicate that sustainability practices enhance businesses' access to capital, as investors prefer companies with strong ESG credentials [2]. Additionally, sustainable projects can build customer loyalty, with consumers favoring brands committed to environmental and social responsibility [3]. Sustainable project management also boosts employee morale and productivity, as employees are more motivated when engaged in projects with positive social and environmental impacts [4].

Sustainable practices improve supply chain efficiency by promoting resource efficiency and reducing waste, resulting in cost savings and enhanced operational performance [5]. Long-term, sustainable projects contribute to better operational performance and cost savings through resource efficiency and reduced environmental impact [6]. Furthermore, sustainable practices enhance shareholder value by improving brand reputation and mitigating risks from environmental regulations and resource scarcity [7].

7.2.1 Financial Advantages

Studies have shown that integrating sustainability practices can improve access to capital for businesses. Investors are increasingly looking to invest in companies with strong ESG practices, making sustainable project management a key factor in attracting investment [2].

7.2.2 Enhanced Customer Loyalty

Consumers are more likely to support brands that demonstrate a commitment to environmental and social responsibility. Sustainable projects that prioritize resource efficiency and ethical sourcing resonate with environmentally conscious customers, leading to increased brand loyalty [3].

7.2.3 *Employee Engagement and Productivity*

Sustainable project management can foster improved employee morale and productivity. When employees are engaged in projects that contribute to positive social and environmental outcomes, they tend to be more motivated and satisfied in their roles [4].

7.2.4 *Long-Term Operational Performance and Increased Shareholder Value*

The long-term benefits of sustainable project management extend beyond immediate project goals. By focusing on resource efficiency and reducing environmental impact, sustainable projects can contribute to improved operational performance and long-term cost savings [6]. Sustainable practices can also create value for shareholders by enhancing brand reputation and mitigating risks associated with environmental regulations and resource scarcity [7].

7.2.5 *Mitigating Business Risks and Talent Acquisition and Retention*

Sustainable project management fosters a proactive approach to risk management, identifying and addressing potential environmental and social risks early in the project life cycle. This can help businesses avoid costly setbacks and disruptions associated with noncompliance with environmental regulations or negative social impacts [8]. Furthermore, a strong sustainability focus can position a company favorably in the public eye, reducing reputational risks associated with environmental or social controversies. Sustainability-focused companies are increasingly seen as attractive employers by talented individuals who prioritize working for organizations that align with their values [9].

7.3 Research Methodology

This chapter employed a literature review methodology, systematically analyzing existing research on sustainable project management and its impact on project outcomes. Academic journals, industry publications, and reports from reputable organizations were used as primary sources. The review focused on research that investigated the connection between implementing sustainability practices within project management and positive business outcomes.

7.4 Discussion

The research reviewed in this chapter overwhelmingly supports the notion that integrating sustainability into project management practices leads to a multitude of positive outcomes. Sustainable projects not only achieve traditional success metrics but also contribute to improved access to capital, customer loyalty, employee morale, supply chain efficiency, and operational performance. Furthermore, sustainable project management practices enhance shareholder value, mitigate business risks, and improve talent acquisition and retention, ultimately contributing to a more resilient and sustainable organization.

7.5 Practical Implications

7.5.1 *Improved Project Outcomes*

By prioritizing resource efficiency, ethical sourcing, and responsible waste management, businesses can achieve cost savings, reduce environmental impact, and enhance operational performance [5].

7.5.2 *Enhanced Brand Reputation and Customer Loyalty*

Sustainable project management practices can help companies build a strong reputation for environmental and social responsibility, leading to increased customer loyalty and brand advocacy [3].

7.5.3 *Risk Mitigation and Improved Resilience*

Proactive identification and mitigation of environmental and social risks within the project life cycle strengthen a company's overall resilience [8].

7.5.4 *Attracting and Retaining Talent*

Companies that embrace sustainable project management principles position themselves as attractive employers for individuals seeking purpose-driven careers [9].

7.5.5 Competitive Advantage

Sustainable project management fosters innovation and efficiency, leading to a competitive advantage in the marketplace [7].

7.6 Conclusion

In conclusion, sustainable project management presents a strategic approach to project delivery, creating positive social, environmental, and economic benefits. By integrating sustainability principles into project management practices, businesses can enhance project outcomes, build a strong reputation, and contribute to a more sustainable future. As the field of sustainable project management continues to evolve, further research is needed to explore its impact on specific industries and project types.

References

1. Project Management Institute (PMI). (2017). *The Standard for Sustainability in Project Management*. Newtown Square, PA: Project Management Institute.
2. Eccles, R. G., Ioannou, I., & Serafeim, G. (2012). The impact of corporate sustainability on organizational processes and performance. *Strategic Management Journal*, 33(8), 1086–1107.
3. Cone Communications. (2017). *Cone Communications Green Millennials 2017 report: A millennial perspective on sustainability*. https://www.cbd.int/doc/case-studies/inc/cs-inc-cone-communications-en.pdf
4. Corporate Responsibility International. (2019). Employee Engagement & Sustainability: A Winning Combination.
5. Carter, C. R., & Ellram, L. M. (1998). Environmental purchasing and firm performance: An empirical investigation. *Journal of Supply Chain Management*, 34(4), 28–35.
6. WBCSD. (2010). *Vision 2050: The new agenda for business*. World Business Council for Sustainable Development. https://www.wbcsd.org/
7. KPMG. (2023). *KPMG Sustainability Reporting Survey 2023*. https://assets.kpmg.com/content/dam/kpmg/se/pdf/komm/2022/Global-Survey-of-Sustainability-Reporting-2022.pdf
8. International Institute for Sustainable Development (IISD). (2020). Embracing Risk Management for Sustainability. https://www.iisd.org/
9. Society for Human Resource Management (SHRM). (2022). *2022 SHRM Talent Management Study*. https://www.shrm.org/

Chapter 8
Driving Sustainable Circularity with Reverse Logistics in Halal Sectors to Degrade Ecological Footprints: Evidence in ASEAN-5

Khairunnisa Musari and Muhammad Rizky Prima Sakti

Abstract This chapter discusses issues related to the circular economy and reverse logistics in the Halal sectors to reduce ecological footprints by taking the case study in the Association of Southeast Asian Nations (ASEAN)-5, namely Malaysia, Indonesia, Singapore, Brunei, and Thailand. All five are among the leaders in global Halal sectors. As a reputable community that engines the international Halal industry, implementing circular economy and reverse logistics in Halal sectors of ASEAN-5 can influence the Halal sectors worldwide, at least in Asia, and other sectors to execute sustainable circularity with reverse logistics to degrade the ecological footprint. Thus, through literature study and netnography, this chapter will explain: (1) Facts and figures of the ecological footprints in ASEAN-5; (2) Facts and figures of the ASEAN-5's Halal sectors; (3) The implementation of the sustainable circularity with reverse logistics in the ASEAN-5's Halal sectors. The findings of this chapter are several implementations in the ASEAN-5 Halal sectors that have begun to be oriented toward sustainable circularity, although most have not yet implemented reverse logistics. Overall, this chapter provides a valuable overview of the implementation of circular economy and/or reverse logistics in the Halal sector in the ASEAN-5 and how these countries have the opportunity to make the Halal industry contribute to reducing the ecological footprints through sustainable circularity with reverse logistics.

Keywords Sustainable supply chains · Sustainability · Circular economy · Halal industry · Logistics · Halal supply chain

K. Musari (✉)
Kiai Haji Achmad Siddiq State Islamic University (UIN KHAS), Jember, Indonesia
e-mail: khairunnisamusari@uinkhas.ac.id

M. R. P. Sakti
University College of Bahrain, Manama, Kingdom of Bahrain
e-mail: mrizky@ucb.edu.bh

B. Jarboui et al. (eds.), *Circular Economy Supply Chains*,
Computational Intelligence Methods and Applications,
https://doi.org/10.1007/978-981-96-7905-8_8

8.1 Background

The Association of Southeast Asian Nations (ASEAN) is one of the 10 top-growing economic regions globally. Referring to ASEAN Statistical Highlights 2023, the economy of this region in 2022 reached a nominal GDP of USD3.6 trillion, positioning itself as the fifth largest in the world and the third largest in Asia, at once became the third largest population in the world [1]. Per August 10, 2024, the United Nations calculated the population in Southeast Asia achieved more than 695 million people, which positions the population rank of this region as number three in Asia or equal to 8.52% of the world's population [2], with the largest population from Indonesia which is sizing to 3.47% of the total world population [3]. ASEAN is presently also one of the pillars of the global Halal sector traffic. Since 2013, five ASEAN member countries have been known as the driving force of the world's Halal industry namely, Malaysia, Indonesia, Brunei Darussalam, Singapore, and Thailand [4–6] with Malaysia's leadership at the forefront of the Halal industry [7], which remains unshaken, having held the top spot for 10 consecutive years [8]. In 2022, Asia is home to 67% of the world's Muslims, and Southeastern Asia is home to 13.8% of Asia's Muslims, after Southern Asia in the first rank as home to 35.6% of Asia's Muslim population [9], which makes ASEAN has a strategical standing in the trans-border Halal sectors commerce. Another side that ASEAN is currently experiencing is the issue of ecological footprint. Malaysia, Indonesia, Brunei Darussalam, Singapore, and Thailand are the top contributors to the ecological footprint in the region [10], even Malaysia, Indonesia, and Thailand are also the 10 top contributors of world mass emissions/ocean plastic waste annually [11]. Considering the leadership of the five in the worldwide Halal market, there is an opportunity for them to be a locomotive as well in the sustainable Halal sectors to degrade ecological footprints. Mainly, in the ASEAN-3, namely Malaysia, Indonesia, and Brunei Darussalam, circularity practices have been implemented in Halal manufacturing [12]. Therefore, through study literature and netnography, this chapter describes the implementation of sustainable circularity with reverse logistics in the Halal industry by taking case studies in Malaysia, Indonesia, Singapore, Brunei Darussalam, and Thailand, where these countries are referred to as ASEAN-5 in this chapter. Both techniques are used to describe: (1) Facts and figures of the ecological footprints in ASEAN-5; (2) Facts and figures of the ASEAN-5's Halal sectors; (3) The implementation of the circular economy and reverse logistics in the ASEAN-5's Halal sectors by providing several examples as case studies that have been implemented. These techniques are taken by considering the unavailability of official publications or reports that specifically present data on the Halal industry that has implemented sustainable circularity and/or reverse logistics in ASEAN-5. Hence, the remaining part of this chapter is organized into five sections. The first section presents the ecological footprints of ASEAN. The Halal sectors in the ASEAN-5 are presented in the second section, while section three explores the implementation of sustainable circularity through reverse logistics in the ASEAN-5's Halal sectors. The subsequent section concludes the entire chapter and provides policy recommendations. In a nutshell, this chapter can be a reference to the implementation of circular economy and reverse logistics in Halal sectors in the leading countries of global Halal by describing their implementations in several Halal sectors in the ASEAN-5.

8.2 Ecological Footprints in ASEAN

Indeed, there is a contradiction between ecological and environmental resilience on the one hand and economic and human development on the other (Musari and Khan, 2023). The friction between economic development and environmental sustainability has long been a warmly debated issue [13]. One embodiment of this matter is the level of ecological footprint following the level of the Human Development Index (HDI) [14]. The higher HDI is followed by an increasingly higher ecological footprint [15, 16], and the ASEAN region is also facing the problem of ecological footprint [10]. In ASEAN, there are not any member states in the Low Human Development category. In this region, there is a tendency that the higher the HDI, the higher the Sustainable Development Goals Index (SDGi), then the higher the GDP per capita, and the higher the ecological [10]. Table 8.1 shows the distribution of socio-demography, economy, HDI level, SDGi achievements, and ecological footprints in the ASEAN Member States. The ecological footprint is indicated by the number of earths required. This table denotes a propensity that the higher the GDP per capita, the higher the HDI, the higher the SDGi, and the higher the ecological footprint.

Furthermore, Table 8.1 also exhibits that the ASEAN-5 are the top six contributors to the ecological deficit and footprints. In the category of very high human development, the member states namely Singapore, Brunei Darussalam, Malaysia, and Thailand consecutively occupy the top ranking for the ecological deficit and the number of earths required (ecological footprints). Then, in the category of high human development, Indonesia follows Vietnam as the member state with the highest ecological deficit and the number of earths required (ecological footprints). Thus, it can be stated that the ASEAN-5, which has the highest HDI and SDGi in the Southeastern Asia region, is strongly indicated to also be a determinant country cluster in producing the ecological footprints in this region.

8.3 Halal Sectors in the ASEAN-5

In the latest state of the Global Islamic Economy Report 2023/2024, the global Islamic economy, also referred to as the Halal lifestyle market, has grown to USD2.29 trillion in 2022, from a USD1.62 trillion consumer spending market in 2012. Based on the Global Islamic Economy (GIE) Indicator, a composite weighted index that measures the overall development of a country's Islamic economic sectors

Table 8.1 Population, GDP, GDP per capita, HDI, SDGi, and ecological footprints in ASEAN member states (2022)

Country	Pop. (000s)	GDP (M USD)	Per capita (USD)	HDI	SDGi	(Deficit)/Res.	Earths
Very high human development							
Singapore	5,673.0	466,602	82,794.6	0.939	71.719	(5.6)	3.8
Brunei	445.4	16,678	37,445.9	0.829	67.100	(5.3)	5.2
Malaysia	32,698.1	407,085	12,467.5	0.803	70.384	(2.2)	2.8
Thailand	66,090.0	495,301	7,494.4	0.800	n.a	(1.2)	1.6
High human development							
Vietnam	99,461.7	408,804	4,110.2	0.726	72.763	(1.6)	1.6
Indonesia	275,719.9	1,317,259	4,777.5	0.713	69.162	(0.5)	1.1
Philippines	111,572.3	404,284	3,623.5	0.710	66.638	(0.8)	0.8
Medium human development							
Laos	7,442.8	15,049	2,022.0	0.620	63.393	(0.2)	1.2
Myanmar	55,770.2	62,990	1,129.5	0.608	64.265	0.2	0.7
Cambodia	16,843.3	29,610	1,758.0	0.600	63.753	(0.3)	0.9

Source: ASEAN Secretariat [1], UNDP [36], Global Footprint Network [37]

by assess- ing their performance in line with broader social obligations, which was launched first in 2013, the ASEAN-5 member states had been and/or always were included in the Top 15 Countries. Until 2023, Malaysia can retain the ranking of 1 for the 10th consecutive year [8]. Table 8.2 shows the ranking for ASEAN-5 within the Top 15 ranking countries based on the GIE Indicator throughout 2013–2023. Referring to this indicator, Thailand has never entered the Top 15. However, in terms of sectors in the Halal industry, Thailand has been in the Top 10 several times, such as in the Muslim-friendly travel, Halal food, and Islamic Finance sectors [5].

Table 8.3 shows the ranking for ASEAN-5 within the Top 10 ranking countries based on the GIE Indicator in 2023 for each Halal sector. Based on this indicator, Thailand in 2023 was among the Top 10 countries with the ranking of 5 in the Halal food sector and ranking of 8 in the Pharmaceuticals and Cosmetics sector. Meanwhile, Malaysia, in addition to consistently occupying the top 15 rankings in the GIE Indicator for 10 consecutive years, also excels in all halal sectors. Since the inception of the GIEI, Malaysia has placed first position. In 2023, Malaysia achieved ranking of 1 in Islamic Finance, Halal Food, and Media and Recreation, with the ranking of 2 in Modest Fashion. In the Pharmaceuticals and Cosmetics sector, four countries within ASEAN-5 were among the Top 10 countries with Singapore in the ranking of 1. Brunei Darussalam in 2023 was not included in Top 10 Countries in any Halal sector.

Certainly, ASEAN-5 has sufficient social capital to strengthen its leadership in the Halal sector. The support of the regional community, namely ASEAN itself, is also a determinant factor. ASEAN can be a best practice in building a Halal-aware community and is capable of promoting the Halal industry in its region at once and has the opportunity to influence other countries in Southeast Asia, the Asia Pacific, or other regions [5]. Besides ASEAN has bilateral trade agreements among member states to open new opportunities for the Halal products trade sector [8], ASEAN also has a number of agreements and institutions that can be legally formal for developing a Halal ecosystem, even for developing a sustainable supply chain ecosystem in the Halal industry [12]. Therefore, the leadership of ASEAN-5 in the global Halal sectors may encourage the ASEAN region to be an anchor for the Halal industrial ecosystem to endorse and reinforce the circular economy [10]. In Southeastern Asia, the circular economy is progressively growing where the companies and the governments present their economic dynamism, political consensus, and industrial innovation to this objective [17]. Further, ASEAN, which includes Islamic finance powerhouses Malaysia and Indonesia, is trying to develop a regional green investments framework, which proves that the ecosystem in this region is seeing several improvements in governance, as well as increased emphasis on green and sustainability issues [18].

Table 8.2 Ranking of the ASEAN-5 in top 15 countries (2013–2023)

ASEAN-5	2013	2014/15	2015/16	2016/17	2017/18	2018/19	2019/20	2020/21	2022	2023
Malaysia	1	1	1	1	1	1	1	1	1	1
Indonesia	5	10	10	10	11	10	5	4	4	3
Singapore	11	13	11	11	13	–	–	15	7	8
Brunei	–	11	–	12	9	11	11	–	–	–
Thailand	–	–	–	–	–	–	–	–	–	–

Source: Thomson Reuters [34], Hidayat and Musari [6, 35], DinarStandard [8, 18]

Table 8.3 ASEAN-5 rankings in 2023

ASEAN-5	Islamic Finan.	Halal food	Muslim Frien. Trav.	Modest Fash.	Media and Recre.	Pharma. and Cosm.
Malaysia	1	1	5	2	1	3
Indonesia	7	2	–	3	6	5
Singapore	–	4	–	4	2	1
Brunei	–	–	–	–	–	–
Thailand	–	5	–	–	–	8

Source: DinarStandard [8]

8.4 Sustainable Circularity and/or Reverse Logistic

Considering ASEAN's leadership in international Halal trade arena [5], the ASEAN-5 has a strong opportunity to become a driver of sustainable circularity in the global Halal sectors, at least in Southeastern Asia, even in Asia. In addition, there is a close relationship between sustainable circularity and Halal sectors. In consonance with the sustainable circularity concept, Halal sectors are also subject to Shari'ah compliance which embraces the obligation to preserve the environment [10]. Referring to the GIE Indicator, there are six sectors of the Islamic economy which also referred to as the Halal economy, namely: (1) Islamic finance; (2) Halal food; (3) Muslim-friendly travel; (4) Modest fashion; (5) Media and recreation, and (6) Halal pharmaceuticals/cosmetics. Another substantial reason for the ASEAN-5 to head sustainable circularity in the global Halal sectors is plastic waste [4, 5, 10, 19], where 6 out of 10 ASEAN member states generating over 31 million tons of plastic waste annually or equally more than half of the world's ocean plastic pollution [11, 20, 21]. Previously, Southeast Asian countries faced a triple whammy of plastic waste, namely their own domestic waste, beached debris, and exported plastic waste from developed countries, where a sufficient infrastructure to manage all of this waste was not available [35]. Further, one of the key enablers of a circular economy is logistics, particularly reverse logistics [12]. Likewise in the Halal industry, one of the key enablers of sustainable circularity is Halal logistics with reverse logistics as the main support. In line with Asia which has a leadership of Halal logistics carried out by Japan [19], the leadership of Halal logistics in ASEAN-5 is also carried out by Japan. Most Halal logistics providers are Japanese companies like Nippon Express, Yusen Logistics, and Kamigumi. The other provider is CXL Halal, a part of Sejung Group, from Korea. Then, one of big challenges to the future of the Halal supply chain, also faced by the Halal logistics, is to develop the closed-loop Halal supply chain to realize a "Circularity of Halal Reverse Logistics" function as a part of the sustainable Halal supply chain, at least by implementing one from 15Rs [12], namely: Refuse (R0), Rethink (R1), Reduce (R2), Reuse (R3), Recycle (R4), Repair (R5), Replace (R6), Replant (R7), Regenerative (R8); Refurbish (R9), Remanufacture (R10), Repurpose (R11), Redesign (R12), Recovery (R13), Remodel (R14), and Reorganization (R15). Therefore, the "Circularity of Halal Reverse Logistics" has become an issue in the Halal sector today, along with the crucial global problems of climate change and its various derivatives that require all parties to contribute to overcoming them. No doubt, sustainable circularity with reverse logistics must be a concern and be realized because this mechanism offers a real solution to contribute to overcoming climate change and the environmental problems it causes. Table 8.4 shows several implementations in the ASEAN-5 Halal sector that have begun to be oriented toward sustainable circularity, although most have not yet implemented reverse logistics. However, it can be predicted that reverse logistics will be the future of sustainable circularity because the two are inseparable things toward "Circularity of Halal Reverse Logistics."

Table 8.4 Sustainable circularity and/or reverse logistics in the ASEAN-5 Halal sectors

Halal sectors	Implementation
Islamic Finance	In **Indonesia**, the government has issued Green Sukuk to develop sustainable transportation network, by upgrading to higher climate resilient design standard and to improve the safety of transporting passengers and logistics, to expand number of travelling schedules for passengers and logistics, and to shifting mode for logistics transportation as reduction of fuel consumption from transportation using trucks. In addition, through Green Sukuk, the government improves the Municipal Solid Waste Management System at Regional and City-scale by developing facilities for applying 3Rs principle (reduce, reuse and recycle), one landfill to generate waste to energy, and leachate treatment [22, 23]. In **Malaysia**, the international financial institutions (IFIs) are spearheading the development and offering of sustainable Islamic trade finance (ITF) products and services under Value-Based Intermediation (VBI). VBI is an initiative introduced by Bank Negara Malaysia (BNM) in collaboration with international financial institutions and aims to deliver Shariah-compliant outcomes through practices, behaviours and offerings that generate positive and sustainable impacts on the economy, society and the environment, and are consistent with sustainable returns and the long-term interests of shareholders [24].
Halal Food	In **Indonesia**, several sugar refinery companies, for example PT Medan Sugar Industry (https://www.msisugar.com) and PT Andalan Furnindo (https://www.afsugar.com), implements halal standard and transform itself to a circular economy by relying on product innovations and appropriate management systems to ensure that the material can remain valuable for a maximum period of time. In addition, in milk industry, milk farmers are moving forward by adopting new technology at milk-collecting points to improve the cold chain process by Halal standard with efficient management of cold chain to confirm the sustainable supply of foods [25]. In **Malaysia**, Unilever (https://www.unileverfoodsolutions.com.my) has collaborated with local councils for recycling the business sustainability among local food operators. In addition, most companies in the livestock industry under Malaysia Productivity Corporation (https://www.mpc.gov.my) have produces poultry, meat, and value-added products with shari'ah-compliant production processes and do reuse wastes, minimize wastes, organic fertilizer, and recycle of nurtrients; In **Brunei**, every producer must ensure that the food, pharmaceuticals, and cosmetics they manufacture are halal and provide the highest standard, which includes ethical and sustainable farming, transportation, slaughter, processing, and logistics to provide integrity throughout the supply chain [26]. In **Singapore**, the charity Free Food for All has pioneered halal ready to eat (RTE) meals that come in retort packaging with flameless heat packs that can be heated without a heat source, also Singapore-based OneAgrix implements a traceability solution for its platform in partnership with Switzerland's INEXTO, a cloud-based global tracking solution [18].

(continued)

Table 8.4 (continued)

Halal sectors	Implementation
	In **Thailand**, Halal Science Centre Chulalongkorn University Chiang Mai Office has launched a series of training courses to empower farmers with innovative technology to stimulate the halal agricultural sector [8]. Likewise, Chula Halal Science Center invented a test kit for manufacturers and consumers capable of detecting five forbidden meat DNA pieces in a single test [18].
Muslim-Friendly Travel	In **Malaysia** and **Indonesia**, they establish themselves as the leading Muslim Among Organization of Islamic Cooperation (OIC) destinations, Malaysia and Indonesia tied for the top spot among 145 destinations, scoring well for ease of entry and quality of tourism infrastructure for both Muslim and non-Muslim travelers. Meanwhile, Singapore has been named the top non-OIC destination for Muslim travelers for the ninth consecutive year, underscoring an unwavering dedication to supporting Muslim travelers with a wide variety of halal food, prayer facilities, suitable airport amenities, and Muslim-friendly accommodation [27]. Women Friendly destination also in OIC Countries with several key factors: their emphasis on safety, absence of faith-based restrictions, and commitment to sustainable socio-economic practices [9]. In **Singapore**, the combination of advanced technology, strong, sustainable tourism initiatives, diverse Halal dining options, and unparalleled safety makes it a top destination in the Global Muslim Travel Index (GMTI). Its forward-thinking approach to embracing technology and sustainability, along with its commitment to catering to the needs of Muslim travelers. Singapore has implemented numerous initiatives that make it a leader in eco-friendly travel. These include extensive green spaces, eco-friendly public transport options, and sustainable practices in hotels and attractions [28].
Modest Fashion	In **Singapore**, modest fashion drives entrepreneurship, inclusivity, airline uniform diversity, sustainable practices, and environmental conservation, for example, fashion designer Hafizah Ghazali plants 100+ trees via IUVA Forest Regeneration program [8]. IUVA Global is a partner in community's environmental and conservation projects in Sumatra and Lombok which has tagline 'People to People, People to Nature' (https://www.instagram.com/iuvaglobal and https://www.facebook.com/IUVA.SG). Designer Hafizah Ghazali believes that beautiful clothes should not only be pretty, iron-free and make women feel good but also not cause negative social and environmental impact [33]. In **Indonesia**, the Indonesian Fashion Chamber (IFC) has made sustainable fashion a key focus of its events by spotlighting the circular economy and create awareness about sustainable fashion among Indonesian youth. The Muslim Fashion Festival (MUFFEST) also featured a sustainability-themed show. Indonesian viscose producer Asia Pacific Rayon (APR) teamed up with designers to create modest fashion collections made from the material [18].
Media and Recreation	In **Indonesia**, there is Nussa, an animated Islamic children's series featuring the main character Nussa with first episode began airing on YouTube in November 2018 and have received over 2 billion views. Then, Nussa & Friends content is available via an app and the Nussa Dunya website, which offers a paid subscription to access content such as related e-books, wallpapers and games. There is also Muslim TV, an Islamic television channel [29].

Pharmaceuticals and Cosmetics	In **Indonesia**, Albéa (https://albea-group.id) produces streamlined compact for refillable eyeshadow palettes of Estée Lauder. Albéa Indonesia serves packaging by integrating the Halal concept and the circular economy paradigm for plastics. Albéa produces 100% of plastic packaging reusable or recyclable, uses a recycled content of 10% from all plastic packaging used, leads action to move on relevant reuse models, and carries action to terminate the unnecessary or tricky plastic packaging. Indeed, Albéa concerns to sustainability for inventing the packaging through Life-Cycle Analysis software and implements plastic weight reduction, eco-design guidelines, development of recyclable packaging, use of bio-based plastics and post-consumer recycled as well as alliances with recycling associations and industries [4, 5]. In **Singapore**, the Government started mandatory packaging reporting in 2022 which requires brand owners, manufacturers, and importers with an annual turnover of more than SGD10 million (US$7.3 million) to file a report on the amount of packaging used annually to the National Environment Agency (NEA) each year. In addition, China's Fujian Haitai Intelligent Technology Co Ltd had its 'Al Noora' disposable face mask halal-certified by United World Halal Development (UNWHD) Singapore, deemed ethical, human, and environmentally friendly [18]. Then, a beauty brand Zahara (https://www.instagram.com/lovefromzahara) offers water-permeable nail polish that are wudhu-friendly, which means they can be worn during prayer as well with the products to not only be Halal-certified, but also vegan and cruelty-free.

Source: Developed by authors

8.5 Conclusion

As the leader in the global Halal economy, the existence of ASEAN-5 plays a major role in influencing the ASEAN region to have high Halal awareness in its member countries and become the pillars of the global Halal sector traffic. On the other hand, ASEAN also faces the problem of ecological footprint with the highest contributor being the ASEAN-5, even 3 of them are also the 10 top contributors of world mass emissions/ocean plastic waste annually. Considering the leadership of the ASEAN-5 in the worldwide Halal market, also as part of the responsibility because of the existence of a high contribution in ecological footprint, there is a must for ASEAN-5 to steer sustainable circularity with reverse logistics in Halal sectors to degrade ecological footprints. Implementing sustainable circularity with reverse logistics in Halal sectors among ASEAN member states, particularly within the ASEAN-5, is a logical choice as a joint movement for this region. Especially, ASEAN has several agreements and institutions that can be legally formal to develop a sustainable Halal ecosystem. It must be admitted that there has been no publication or official report that specifically describes the Halal industry that has implemented sustainable circularity and/or reverse logistics in ASEAN-5, especially to prove that this practice can significantly reduce the ecological footprint in the ASEAN region. For that, of course, further research is needed or adequate data is available to be able to detect it. However, conceptually, sustainable circularity—or what is often called a circular economy—with reverse logistics offers a real solution to contribute to degrading ecological footprints, which in turn will overcome climate change and the environmental problems it causes. Overall, the findings of this chapter are several implementations in the ASEAN-5 Halal sectors that have begun to be oriented toward sustainable circularity, although most have not yet implemented reverse logistics. The precious contribution of this chapter is the encouragement for countries with high ecological footprints to be responsible in their leading industries by implementing, among others, circular economy and/or reverse logistics. This chapter provides a worthwhile overview of the implementation of circular economy and/or reverse logistics in the Halal sector in the ASEAN-5 and how these countries have the opportunity to make the Halal industry contribute to reducing the ecological footprint through sustainable circularity with reverse logistics.

8.6 Policy Recommendation

This chapter provides two recommendations for driving sustainable circularity with reverse logistics in Halal sectors to degrade ecological footprints by the ASEAN-5. First, encourage the implementation of sustainable circularity and reverse logistics, particularly for waste material, as a requirement in the Halal industry and one of the main pillars in the ASEAN Halal ecosystem. ASEAN can set a sustainable Halal supply chain roadmap, including implementing the circularity 15Rs in the

Halal industry and developing its reverse logistics infrastructure, to address end-to-end supply chain transformation and, in turn, realize a sustainable Halal supply chain. It can be a comparative advantage for the ASEAN Halal economy and at once leads radically to reducing plastic pollution in the Halal market. The leadership of ASEAN with the ASEAN-5 as the engine has opportunities for influencing this value to the Halal market in Asia, even globally.

Second, encourage ASEAN through the ASEAN Secretariat to be more active in voicing the enforcement and implementation of regulations to ASEAN member countries to implement sustainable circularity and reverse logistics in the industrial ecosystem and foreign trade. This is considering that in reality all ASEAN member countries produce ecological footprints, and 6 of 10 ASEAN Member States are the Top 10 Countries of world mass emissions/ocean plastic waste annually, equally more than half of the world's ocean plastic pollution.

References

1. ASEAN Secretariat: ASEAN statistical highlights 2023. Jakarta: The Association of Southeast Asian Nations (ASEAN) Secretariat.
2. https://www.worldometers.info/world-population/indonesia-population/
3. Worldometers. (202b). South-Eastern Asia population (per August 10th, 2024). https://www.worldometers.info/world-population/south-eastern-asia-population/
4. Musari, K.: Circular economy for plastics and digitally-enabled community towards ASEAN halal hub in Asia. In: de Pablos, P.O. (Ed.), Handbook of research on developing circular, digital, and green economies in Asia, pp. 1-12. IGI Global, Hershey, PA (2021) doi:10.4018/978-1-7998-8678-5.ch001
5. Musari, K.: Driving sustainable supply chain and digital transformation in Halal sectors: evidence in ASEAN-5. In de Pablos, P.O., Almunawar, M.N., Anshari, M. (Eds.), Strengthening sustainable digitalization of Asian economy and society, pp. 87-100. IGI Global, Hershey, PA (2024a) https://doi.org/10.4018/979-8-3693-1942-0.ch005
6. Hidayat, S. E., Musari, K.: ASEAN towards a global Halal logistics through the digitally-enabled community. International Journal of Asian Business and Information Management (IJABIM), 13(2), 1-15 (2022) https://doi.org/10.4018/IJABIM.20220701.oa1
7. Abdullah, M.A., & Azam, Md.S.E.: Halal industry in ASEAN: Issues and challenges. In: de Pablos, P.O., Almunawar, M.N., Abduh, M. (Eds.), Economics, business, and Islamic finance in ASEAN economics community, pp. 77-104. IGI Global, Hershey, PA (2020)
8. DinarStandard: State of the global Islamic economy report 2023/2024. Dinar- Standard, Salaam Gateway & Dubai Economy and Tourism, Dubai (2023)
9. Mastercard-CrescentRating: Global Muslim travel index 2022. CrescentRating Pte. Ltd. and Mastercard Asia/Pacific Pte Ltd., Singapore (2022)
10. Musari, K., Sakti, M.R.P.: Recycled waste plastic for asphalt manufacturing, can it degrade ecological footprints in ASEAN? A policy analysis from the Islamic Economic Approach. In: Sakalayen, Q.M.H., Almunawar, M.N., Ali, M.Y. (Eds.). Sustainable advanced manufacturing and logistics in ASEAN, pp. 269–288. IGI Global, Hershey, PA (2024) https://doi.org/10.4018/979-8-3693-5350-9
11. Meijer, L.J.J., van Emmerik, T., van der Ent, R., Schmidt, C., Lebreton, L.: More than 1000 rivers account for % of global riverine plastic emissions into the ocean. Science Advances (2021) https://doi.org/10.1126/sciadv.aaz5803

12. Musari, K., Mahmudah, Hakim, Z., Almunawar, M.N.: Mapping the implementation of circular economy and reverse logistics in the sustainable halal supply chain: Evidence in ASEAN-3. In: Sakalayen, Q.M.H., Almunawar, M.N., Ali, M.Y. (Eds.). Sustainable advanced manufacturing and logistics in ASEAN, pp. 61–78. IGI Global, Hershey, PA (2024) https://doi.org/10.4018/979-8-3693-5350-9
13. ElMassah, S., & Hassanein, E.A. Economic development and environmental sustainability in the GCC countries: new insights based on the economic complexity. Sustainability 15, 7987 (2023) https://doi.org/10.3390/su15107987
14. Musari, K.: Transforming ecological footprint through circular economy innovation towards sustainable well-being in the development of (Islamic) smart cities in GCC countries. In: Lytras, M.D., Alkhaldi, A., Malik, S. (Eds.), The emerald handbook of smart cities in the gulf region: innovation, development, transformation, and prosperity for vision 2040, pp. 379–403. Emerald Publishing Limited, Bingley, UK. (2024b)
15. Khan, T.: Rethinking economics in the anthropogenic era. A presentation material for Guest Lecture at Quaid-i-Azam University, Islamabad, October 13th (2022)
16. Khan, T.: Externalities management from Islamic economic perspective: rethinking economics and finance in the anthropogenic era. A presentation material for Kulliyyah of Economics & Management Sciences (KENMS) of International Islamic University Malaysia (IIUM) Distinguished Intellectual Discourse Series, A Prelude to Key Themes in the 15th International Conference on Islamic Economics & Finance (ICIEF) 2023 Conference, May 26th (2023)
17. ERIA: Integrative report on implementation of the circular economy in ASEAN. ERIA Research Project Report FY2023 No. 27. Economic Research Institute for ASEAN and East Asia (ERIA), Jakarta (2023)
18. DinarStandard: State of the global Islamic economy report 2022. DinarStandard, Dubai Islamic Economic Development Center, and Salam Gateway, Dubai (2022)
19. Musari, K., & Zaroni: Reverse logistics in the age of digital transformation for circular economy and Halal logistics through the leadership of Asia. In: de Pablos, P.O., Zhang, X., Almunawar, M.N. (Eds.), Handbook of research on disruptive innovation and digital transformation in Asia, pp. 83-103. IGI Global, Hershey, PA (2021). https://doi.org/10.4018/978-1-7998-6477-6.ch006
20. Zorzi, L.: How to defeat the plastic tide threatening the ASEAN region's green growth. World Economic Forum, (2023, October 10). https://www.weforum.org/agenda/2023/10/defeat-plastic-tide-threatening-asean-green-growth
21. Wicaksono, L.L.: Which countries pollute the most ocean plastic waste? Visual Capitalist (2023, February 17). https://www.visualcapitalist.com/cp/visualized-ocean-plastic-waste-pollution-by-country
22. MoF: Green sukuk, allocation and impact report 2022. Ministry of Finance (MoF) Republic of Indonesia, Jakarta (2022)
23. MoF: Green sukuk, allocation and impact report 2023. Ministry of Finance (MoF) Republic of Indonesia, Jakarta (2023)
24. World Bank: Islamic trade finance, an opportunity for Malaysia. International Bank for Reconstruction and Development/The World Bank, Kuala Lumpur (2021)
25. Mulyaningsih, H. D., Arifin, R. W. D., Nu'man, A. H., Rachmawati, A. W., and Rahmawati, S.: Halal supply chain and halal logistics of the cold chain model in the milk industry: evidence from milk cooperatives in Indonesia. In: Rahman, N.A.A., Hassan, A., Mohammad, M.FN. (Eds.), Halal logistics and supply chain management in Southeast Asia, pp. 21-36. Routledge, Oxon New York (2020)
26. Rahim, S.M., Sulaiman, N.R.: Role of Industry Players in Brunei's Halal Industry Development. Social and Humaniora Research Symposium (SoRes), KnE Life Sciences, 1030–1044 (2023). https://doi.org/10.18502/kss.v8i18.14309
27. Chew, N.: The top Muslim travel destinations in the world. Travel Weekly Asia (2024, June 10). https://www.travelweekly-asia.com/Travel-News/Travel-Trends/The-top-Muslim-travel-destinations-in-the-world
28. Mastercard-CrescentRating: Global Muslim travel index 2024. Singapore: CrescentRating Pte. Ltd. and Mastercard Asia/Pacific Pte Ltd., Singapore (2024)

29. DinarStandard, Indonesia Halal Lifestyle Center, Bank Indonesia: Indonesia halal markets report 2021/2022. Jakarta: DinarStandard, Indonesia Halal Lifestyle Center, and Bank Indonesia, Jakarta (2021).
30. Fennemann, V., Hohaus, C., & Kopka, J-P.: Moving in circles: Logistics as key enabler for a circular economy. Whitepaper Fraunhofer Institute for Material Flow and Logistics, 9, July (2018). https://doi.org/10.24406/IML-N-502288.
31. Queiroz, M. M., Pereira, S. C. F., Telles, R., & Machado, M. C.: Industry 4.0 and digital supply chain capabilities: A framework for understanding digitalisation challenges and opportunities. Benchmarking: An International Journal (2019) https://doi.org/10.1108/BIJ-12-2018-0435
32. Gottlieb, S., Ivanov, D., Das, A.: Case studies of the digital technology impacts on supply chain disruption risk management. In: Schro¨der, M., Wegner, K. (Eds.). Logistik im wandel der zeit – von der production ssteuerung zu vernetzten supply chains, pp. 23- 52. ein Teil von Springer Nature, Gabler Verlag Springer Fachmedien Wiesbaden GmbH (2019) https://doi.org/10.1007/978-3-658-25412-42
33. Abdullah, A. F. B.: This Muslim fashion designer has planted over 100 trees. Muslim.Sg, (2020, January 6). https://muslim.sg/articles/this-muslim-fashion-designer-has-planted-over-100-trees
34. Thomson Reuters: State of the global Islamic economy 2013 report. Thomson Reuters, DinarStandard™ & Dubai The Capital of Islamic Economy (2013).
35. Hidayat, S. E., Musari, K.: Comparative analysis of digitally-enabled community in supporting the Halal industry in Muslim-majority countries in the ASEAN, Central Asia, and Maghreb region. In: Rafiki, A. (Ed.), Digitalization in halal management. Contributions to Environmental Sciences & Innovative Business Technology, pp. 151-165. Springer, Singapore (2023). https://doi.org/10.1007/978-981-99-5146-8-11.
36. UNDP: The 2023/2024 human development report. United Nations Development Programme (UNDP) (2024). https://hdr.undp.org/data-center/human-development-index
37. Global Footprint Network: 2023 edition national footprint and biocapacity accounts. (2023). https://www.footprintnetwork.org/licenses/public-data-package-free/

Chapter 9
Circular Logistics and Operations: Transportation Across a Product's Lifecycle

Emrah Demir, Thanos E Goltsos, and Sean White

Abstract The transition to the circular economy is underway but needs to overcome decades of linear thinking underpinning operations optimization and practices. The main difference between linear and circular operations is the role of consumers, from just recipients of finished goods to also being suppliers of cores (used products), which repair centers, remanufacturers, and other CE stakeholders can use as inputs to their operations. This transformation or change of roles underpins a huge increase of uncertainty, as supply is no longer limited in a mostly business-to-business-to-customer context, but rather customer-to-business-to-customer context, and has important ramifications on all operations, including transportation. With the goal being to eliminate materials reaching landfill, it is reasonable to expect that any forward (delivery) operation needs to be mirrored by a reverse (collection) operation, to close the loop and support circular logistics. This change is not trivial nor straightforward, as it is also not properly understood. To address this complex but crucial problem, this chapter proposes numerical analysis for designing a data-driven, operation-level decision-making support system that can dynamically allocate limited resources (e.g., transport capacity) to the right operations. We apply Monte Carlo simulation to consider pre-inspection, consolidation, and a product's lifecycle for the chosen real-life case study. We employ forecasting to determine quality-based allocation across CE stakeholders (repair, remanufacturing, recycling) and consider various factors on a synthesized dataset, empirically grounded on data and discussions from and with experienced managers in the area. We find that pre-inspection and consolidation at the retailer is beneficial and that the benefits depend on the warranty duration relative to customer behavior (demand and return rates).

E. Demir (✉) · T. E. Goltsos
Logistics and Operations Management Department, Cardiff Business School, Cardiff University, Cardiff, Wales, UK
e-mail: demire@cardiff.ac.uk; goltsosa@cardiff.ac.uk

S. White
Global Head of Repair Services, DSV; PARC Institute of Manufacturing, Logistics and Inventory, Cardiff Business, School, Cardiff University, Cardiff, Wales, UK

B. Jarboui et al. (eds.), *Circular Economy Supply Chains*,
Computational Intelligence Methods and Applications,
https://doi.org/10.1007/978-981-96-7905-8_9

Keywords Circular economy · Transportation · Product lifecycle · Inspection · Monte Carlo simulation

9.1 Introduction

Until very recently, the vast majority of supply chains and economic models operated in a "linear" fashion. This traditional economic model describes the typical operation in which raw materials are extracted, transformed into products and parts, sold, used, and discarded, often coined as a "make-use-dispose" linear economy. This operating model is not sustainable and needs to be superseded by "circular economic" models if we are ever to achieve social, environmental, and economic sustainability [13]. A circular economy (CE) is then an alternative to a traditional linear economy in which we keep materials, parts, and products in use for as long as possible, extract the maximum value from them while in use, and then recover and regenerate products and materials at the end of each service life (as opposed to their end of life, in linear economic thinking). The transition to a circular economy is integral to achieving the UN's Sustainable Development Goals (SDG) [16].

In the CE, products are designed with subsequent circular economic treatments (repair, remanufacture, repurpose, etc.) in mind. These are collectively discussed as circular economic "strategies," each of which categorically describe different treatment options [7, 14]. The relevance of the strategies to a particular supply chain or product depends on the business model, its design and general product characteristics, as well as product lifecycle (PLC) considerations [26]. This means that to be able to fully capture the value of the returned products (cores), a CE strategy mix has to be well thought out and implemented and coordinate different stakeholders (manufacturers, retailers, remanufacturers, recyclers, etc.).

In the CE, materials are circulated within and between different supply chains and markets and are often repaired or transformed. This is enabled by customers returning products to CE stakeholders at their end of use [17]. This in turn means that customers are transformed from mere recipients and users of products to also become suppliers to CE stakeholders. This transformation or change of roles underpins a huge increase of uncertainty, as supply is no longer limited in a mostly business-to-business-to-customer context, but rather customer-to-business-to-customer context, and has important ramifications on all operations, including transportation [18].

It is for the above reasons that the logistics of circular supply chains (CSC) face several challenges, such as poor predictability of product, part, and material streams (in terms of both rate and quality/condition) and variation of the returns and their quality. Forecasting can help in this direction by creating estimates of demand and return rates, and it has however been reported as an area in need of contributions.

Solutions include a shift in business models (e.g., retaining product ownership and leasing products to customers) and design principles [19]. As circular logistics (CL) increases in importance and companies and consumers switch from, e.g., ownership of products to the access of products as a service [20, 21], finding logistics solutions becomes more relevant and pressing.

Cost-efficient and ecological supply chain management is a precondition for a well-functioning CE. Collaboration between diverse companies and supply chains is needed to bring costs down and attain economies of scale. This includes collaboration between repair center operators, remanufacturers, repurposers and recyclers, among others. In many cases, shortcomings in coordination form obstacles for the transition or at a minimum impair its profitability potential.

The collection channel is an important consideration when designing the CL and the CSC that contribute to it, as it affects the transportation plans for both delivering to retailers and customers and intercepting products which would otherwise end up in landfill [10]. Options include the original equipment manufacturer (OEM), CE stakeholder (recycler, remanufacture), or 3rd party (core broker), or combinations of the above [17]). An equally important parallel consideration is about how this first point of contact of the reverse loop (CL return channel) might add value to the operation. Pre-inspection (or presorting as it is often referred to in the literature) is important as it enables diversification of returns based on quality (condition) considerations [23]. Based on the above, the CL model supply chain can be seen in Fig. 9.1.

We consider a CL supply chain consisting of customers, a retailer, and a repair/remanufacturing center (which can do both repairs and remanufacturing of cores), as well as recycling centers. We distinguish between warranty returns which need to be repaired within a predefined time period and out-of-warranty returns which will be remanufactured. Returns that are beyond economical repair will eventually find their way to a recycling facility. While warranty repairs are all funneled to the repair operation, if only to be found void (for beyond

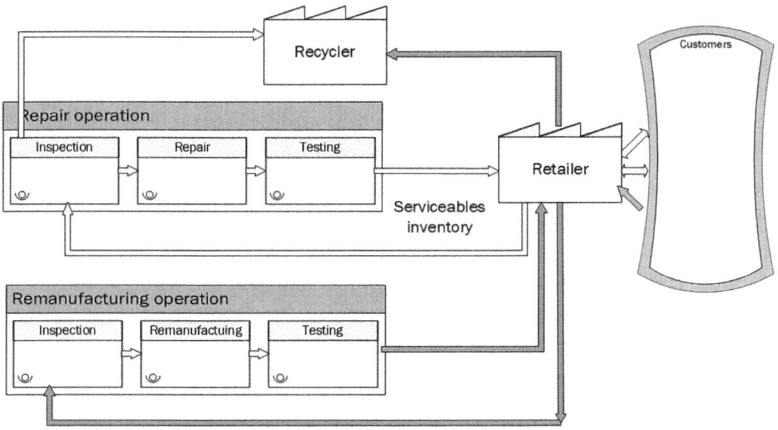

Fig. 9.1 A stylized CL supply chain consisting of a retailer, a remanufacturing/repair operator, and a recycler (white arrows signify warranty returns, while gray arrows out-of-warranty returns)

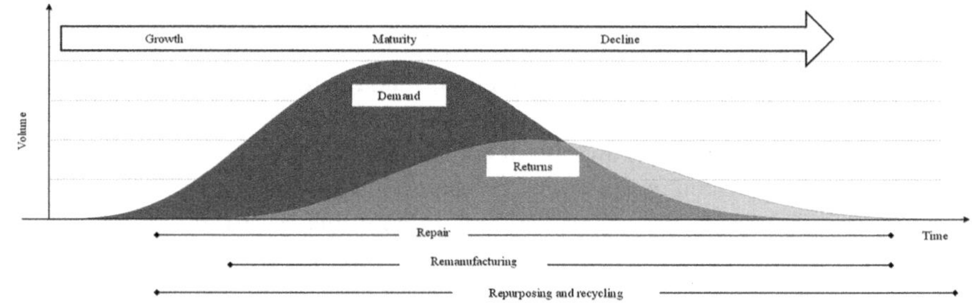

Fig. 9.2 A product lifecycle incorporating product returns and their condition

economical repair returns), the out-of-warranty remanufacturing depends on the quality (condition) of the returns and the position in the product lifecycle.

The product lifecycle then becomes a very important consideration in CL as it affects the relevance of different CE strategies as a product moves from its introduction through its decline. By incorporating product lifecycle considerations, businesses can design product processes and CL supply chains, whose design is conductive to a contextualized CE strategy "mix," therefore reducing waste and extending meaningful product, part, and material life. This approach allows for environmental impact by closing the loop, transforming end-of-life output waste (in the linear economy sense) into input resource for extending meaningful life or new production.

In this research, we particularly focus on a product's lifecycle to consider data-driven, intelligent consolidation for effective transportation planning. Figure 9.2 shows the product's lifecycle and various stages. We note that we use PLC with relation to the demand of the product, but the PLC could be redefined or adapted to describe either the repair or the remanufacturing "life-cycles." However, and as explained later, we employ forecasting of demand and out-of-warranty returns to distinguish between remanufacturing-related phases, and how those impact on the transportation of cores from the retailer.

The contributions of this chapter are four-fold. First, we investigate various CE strategies, namely: repair, remanufacturing, and recycling, and their effect on CL. Second, we contextualize operations against the product's lifecycle. To the best of our knowledge, these strategies are rarely considered together from a CL perspective, considering product and return lifecycles. Third, we engage with the fundamental concepts of pre-inspection and consolidation, and product's lifecycle for returns, and investigate their impact in the transportation plans. Fourth, we consider real-life data from an international electronics company and adapt their operations to the ones described in Fig. 9.1. The unique data has allowed us to investigate various scenarios to better understand the process of returns, as well as investigate the impact of warranty length. Finally, we provide several insights to further optimize return CL operations.

The content of this chapter is structured as follows. We first provide a brief literature review on recent and related circular economy studies focusing on circular economy strategies and transportation in Sect. 9.2. Section 9.3 provides the formal definition and mathematical foundations of the studied CL supply chain model. In the following section, we provide the details of the case study and present numerical results to compare several key performance indicators (KPIs). Section 9.5 provides the discussion, conclusions, and future research avenues.

9.2 Literature Review

This section provides a brief review of the literature relevant to our work, namely CE strategies, product lifecycle in a CL context, return forecasting, and the role of transportation for CE transformation.

9.2.1 Circular Economic Strategies

In the CE, there are several strategies that influence the configuration of the logistic network, including recycling, remanufacturing, and repair, which we engage with in the current research.

Recycling involves breaking down used materials to create new products, reducing the need for virgin resources. Recycling is often discussed as "down cycling" as it (i) loses all the engineered value of a product and (ii) often has a material output of lower quality than the input. Remanufacturing goes further by restoring a product to like-new condition, most often involving disassembly, inspection, replacement of parts, and testing, allowing for continued use of a product in its original function. Remanufacturing is closely associated with the assumption of "perfect substitution," which means that a remanufactured product is indistinguishable from a new product, in terms of both use and customer perception. We adopt this assumption. Finally, repair includes maintenance and/or repairing a product so it can fulfill its initial function [29]. It can be within warranty or outside, and while conceptually very similar to remanufacturing, it involves "lighter-touch" interventions and does not typically renew any warranties.

CL plays an essential role in CE systems. Circular strategies such as repair and remanufacturing not only reduce the demand for new products but also shorten product supply chains while extending meaningful life of products and parts. Blomsma and Brennan [8] provide evidence that efficient product return systems, where products are returned for remanufacturing or recycling, can drive the economic viability of the CE. By maximizing the lifespan of materials, parts, and products, these strategies reduce the overall environmental footprint associated with transportation and logistics.

9.2.2 Forecasting and Inventory Control for the Circular Economy

Forecasting and inventory control play an important role in the effective management of supply chains, especially within closed-loop systems where demand and returns are intertwined [18, 28]. In these systems, it is essential to forecast not only demand, but also the return rates, as both significantly influence operational performance. Remanufacturers and other CE stakeholders face unique challenges related to supply uncertainty, particularly in the reverse loop, where the forecasting of returns becomes less accurate due to the complexity of modeling diverse customer (i.e., supplier, in the reverse logistics context) behavior.

Within CL, two primary areas of supply uncertainty need forecasting: the rate and quantity of returns, and the quality (condition) of those returns [27] (on top of the rate of demand). The timing of when a return will occur and the condition of returned items are both vital factors that impact the remanufacturing (or repair) process. While the rate of returns has received some attention, the quality facet remains largely unexplored, despite its significance, perhaps hindered by how hard it is to forecast. It is for this and other reasons that presorting cores through inspection, before these reach the repair or remanufacturing stages, is very important.

The use of digitalization and data analytics in CE also supports transportation efficiency. Digital platforms enhance transparency, enabling organizations to track the location, condition, and availability of assets and products within the CL system, which could at the extremes supersede the need for presorting and/or forecasting [9].

In our case, we employ simple forecasting methods as forward signaling to define which quality categories are to be sent to repair/remanufacturing or to recycling (and how).

9.2.3 Structuring the Collection Channel and Presorting of Cores

The selection of a collection channel and its capabilities (e.g., consolidation, pre-inspection) can alleviate some of the uncertainty stemming from inaccurate rate and quality return forecasts. There are three main channels for collecting used products: (i) collection by the manufacturer, (ii) by the retailer, or (iii) by a third party, such as specialized collection companies or nonprofits known as "core brokers" (e.g., [22]). Collection can occur through either a single channel—using only one of the above—or a multichannel approach that combines two or more. Savaskan et al. [10] showed that in a decentralized system led by the manufacturer, retailers were most efficient for collection due to their proximity to customers. Atasu et al. [11] also agreed that retailer-led collection is best suited for economies of scale. Choi et al. [12] analyzed closed-loop supply chain efficiency under varying channel leadership and found the retailer-led collection optimal for environmental and societal welfare. In our study, we chose to use the retailer as the single point of (re-)entry to the circular supply chain.

With regard to inspection, [23] studied Pitney Bowes' remanufacturing operations and recommended that presorting returns upon arrival lead to 4%+ cost savings compared to no sorting. This approach aligns with [24], who showed that quality-based sorting can cut costs, particularly when returns exceed demand or demand is slow-moving (as is the case in the later stages of PLC, see Sect. 9.4). In another study, [25] further demonstrated that even basic (pre-)sorting can provide essential quality insights and improvements for remanufacturing. While these studies focus on the benefits of presorting on the remanufacturing operations (or indeed, any other circular operations strategy such as repair), we will show how pre-inspection at the retailer level also brings transportation-related benefits on CL.

The literature on core collection underscores the need to manage uncertainties in return rate and quality. Collection processes can affect both the volume and quality of returns between stakeholders, making the selection of channels and their capabilities important for managing supply uncertainty in CL. Despite its importance, closed-loop models often assume fixed return rates and uniform quality for simplicity, which overlooks the impact of customer behavior on the supply chain and the actual variability in return quality and quantity.

9.2.4 Circular Logistics and Transportation

Several prominent studies highlight the role of transport in CE [1], including optimizing logistics operations for waste reduction, consolidation, and implementing reverse logistics systems in which products are circularly treated instead of disposed of. Such strategies align with the key UN (SDG) and CE goals of minimizing resource consumption and reducing their environmental impact [2].

CL emphasizes reverse logistics, which encompasses activities such as collection, sorting, consolidation, and redistribution of used products (cores). It enables efficient resource utilization and ensures that returned products, parts, or materials reenter the supply chain at the appropriate point, often based on considerations around lifecycle and the condition of the product. In that way, logistics providers or retailers can add further value to the supply chain, providing necessary support to the CSC. Govindan and Soleimani [3] demonstrate how reverse logistics improve sustainability, showing that reintroducing used products back into the supply chain reduces raw material demand and waste. Moreover, [4] indicate that CE-based transportation systems need adaptive supply chain designs that respond to diverse customer and product needs, which is especially relevant in multimodal transport networks. In these systems, CE strategies help reduce greenhouse gas emissions and energy use by optimizing transport routes and selecting eco-friendly transportation modes, such as rail or maritime over road transport for longer distances [5].

Another key area in CE transportation strategies is related to vehicle utilization and transport optimization. One of the primary challenges of CE is achieving efficient transport operations without significant cost increases. Stock and Boyer [6] points out how transport optimization can be considered in CE systems, where digital tools, such as route optimization software and AI-based forecasting models, can predict demand and returns more accurately, reducing the need for redundant empty trips and improving payload factors. Studies by Kirchherr et al. [15] add that efficient vehicle utilization not only reduces emissions but also extends vehicles' lifecycles, thus promoting asset circularity. Also, it is important to note that effective use of available vehicle space is a prerequisite for CL.

For example, full-truckload (FTL) over less-than-truckload (LTL) services can lead to a reduction in transport costs and "empty miles" by promoting consolidation and reducing total distance [30]. In this research, we specifically focus on LTL and FTL shipments as transport services coordinated by a retailer via third-party logistics company to highlight the importance of consolidation for CL supply chains.

9.3 Problem Description and Modeling

This section formally defines the problem at hand and provides the mathematical foundations. We consider a supply chain network consisting of four entities: customers, a retailer, a repair-remanufacturing facility, and a recycler. The process begins with customers sending their products to retailers. The reason for this might be a product fault, or just returning it at their end of use. Some of these products are within warranty and need to be repaired, within 2 weeks. To improve the product return process, retailers may perform a pre-inspection to categorize the products into four categories: great (α), good (β), bad (γ), and beyond economical repair (BER) (δ). Due to the possibility of misclassification, there is a probability (p_{mis}) that the retailer miscategorizes the returned products. This refers to a $(p_{mis})/2$ chance that the product is mischaracterized either as better than it is or worse. We note that if the product's actual condition is either great or BER, there is a p_{mis} chance that the product will be mischaracterized as good or bad, respectively.

The transport shipment plans according to the position on the product's lifecycle (i.e., growth, maturity, decline) are decided as shown in Table 9.1. If the returned product is within warranty, great (α), good (β), and bad (γ) are sent directly to remanufacturers as LTL shipments. In the case where pre-inspection is in place, BER products (δ) are sent directly to the repair facility in batches as FTL, or as LTL (along with everything else) when there is no pre-inspection.

Upon receiving the products, the remanufacturers perform a so-called actual inspection as part of the repair or remanufacturing operation, where the actual condition of the product is revealed, and the products are reclassified. For warranty returns, α, β, and γ products are repaired and sent back to the retailer for distribution (LTL), while δ products are sent to recyclers (FTL).

For out-of-warranty returns, everything that is sent to the remanufacturer from the retailer is remanufactured. What is received is defined by forecasts of demand

Table 9.1 Circular logistics flows based on product condition and product lifecycle consideration

	Growth		Maturity		Decline	
	Condition	Shipment	Condition	Shipment	Condition	Shipment
From RET To REM						
Within warranty	(α), (β), (γ)	LTL	(α), (β), (γ)	LTL	(α), (β), (γ)	LTL
	(δ)	FTL	(δ)	FTL	(δ)	FTL
Out of warranty	(α), (β), (γ)	LTL	(α), (β)	LTL	(α)	LTL
From RET to REC						
Out of warranty	(δ)	FTL	(γ), (δ)	FTL	(β), (γ), (δ)	FTL
From REM to RET						
Within warranty	(α), (β), (γ)	LTL	(α), (β), (γ)	LTL	(α), (β), (γ)	LTL
Out of warranty	(α), (β), (γ)	LTL	(α), (β)	LTL	(α)	LTL
From REM to REC						
Within warranty	(δ)	FTL	(δ)	FTL	(δ)	FTL
From RET to C						
Within warranty	(α), (β), (γ)	LTL	(α), (β), (γ)	LTL	(α), (β), (γ)	LTL

RET: retailer, REM: repair-remanufacturing facility, C: customer, REC: recycler
(α): great; (β): good; (γ): bad, (δ): beyond economical repair
LTL: less than truckload, FTL: full truckload

and out-of-warranty repairs per quality class, which define the stage of the PLC (see Table 9.1, Algorithm 2, and Fig. 9.4b, c, and d). For example, when the forecast of α quality out-of-warranty returns is greater than the forecast of demand, only α quality products are sent to the remanufacturer (LTL), while the rest are sent to the recycler (FTL). (As remanufacturing that category alone is enough to cover demand.) If on the other hand that forecast is less than the demand, while the joint forecast of α and β out-of-warranty returns is greater than the demand forecast, α and β products are sent to the remanufacturer (LTL), while γ and δ to the recycler (FTL).

The objective is to model the flow of products through this CL supply chain network, considering condition of returns, demand in relation to returns, the probability of misclassification and reclassification, and to optimize the whole process using Monte Carlo simulation.

9.3.1 *Mathematical Formulation*

We now provide the mathematical foundations of the studied operational-level CL problem. We first define the sets used in Table 9.2.

We note that remanufacturers can do both repair and remanufacturing. We provide the list of parameters and decision variables used for the problem in Table 9.3.

The flow of products under warranty can be described by the following equations. We first present the situation when there is a pre-inspection by a retailer as follows

Algorithm 1 The product flow in the circular logistics problem

Input: Faulty products from customers and demand for new products
Output: Repaired products, remanufactured products, recycled products
procedure CUSTOMEROPERATIONS
    Customers send faulty products to the retailer
end procedure
procedure PREINSPECTIONBYRETAILER
    Retailers categorize products into $\alpha, \beta, \gamma, \delta$
    With probability ($p_{\text{mis}} = 0.25, 0.50$), products are miscategorized
end procedure
procedure FLOWPOSTPREINSPECTION
    if Product is within its warranty. **then**
        if Inspected condition is α, β, γ **then**
            α, β, γ products are sent to REM via LTL,
        else
            δ products categorized as BER are sent to REM via FTL
        end if
    else Out of warranty
        Algorithm 2 defines the position at the product lifecycle and the appropriate set of items are sent to REM via LTL and the rest to REC via FTL.
    end if
end procedure
procedure REMANUFACTUREROPERATIONS
    REM inspects products
    Products are correctly reclassified
    δ products are sent to REC
    if Product is within warranty **then**
        if Condition is α, β, or γ **then**
            Products are repaired and sent back to RET via LTL
        else
            Products are sent to REC via FTL
        end if
    else Out-of-warranty Remanufacture all arrivals (as decided by Algorithm 2)
    end if
end procedure
procedure RECYCLINGOPERATIONS
    Recycle all arrivals
end procedure

Table 9.2 Notation and description of sets

Notation	Description
C	Set of customers $C = \{1, ..., n\}$
R	Set of retailers $R = \{1\}$
M	Set of remanufacturers $M = \{1\}$
L	Set of recyclers $L = \{1\}$
Q	Set of conditions $Q = \{\alpha, \beta, \gamma, \delta\}$

(for both within and outside warranty).

$$y_\alpha = (1 - p_{\text{mis}}) \cdot Z_\alpha + (p_{\text{mis}})/2 \cdot (Z_\beta)$$

Table 9.3 Parameters and decision variables

Parameter/Variable	Description
p_{mis}	Probability of misclassification by retailer (i.e., $p_{\text{mis}} = 0.25$ or $p_{\text{mis}} = 0.50$)
d_{ij}	Distance between i and j (i.e., $i \in \{C, R, M\}$, $j \in \{C, R, M, L\}$)
x_{ij}	Number of products sent from entity i to entity j
y_k	Number of products in category k after pre-inspection ($k \in \{\alpha, \beta, \gamma, \delta\}$)
z_k	Number of products in category k after remanufacturer inspection

Algorithm 2 The position of product lifecycle

```
1: Input: Demand, Our-of-warranty returns
2: Output: Product lifecycle position
3: procedure RETAILERTOREMANUFACTURER
4:     if Aggregate forecast of α, β, and γ out-of-warranty returns < forecast of demand then
5:         Growth phase: remanufacture α, β, and γ, recycle δ
6:     else if Aggregate forecast of α and β out-of-warranty returns > forecast of demand then
7:         Maturity phase: remanufacture α, β, recycle γ and δ
8:     else if Forecast of α out-of-warranty returns > forecast of demand then
9:         Decline phase: remanufacture α, recycle α, γ, and δ
10:    end if
11: end procedure
```

$$y_\beta = (1 - p_{\text{mis}}) \cdot Z_\beta + (p_{\text{mis}}) \cdot Z_\alpha + (p_{\text{mis}})/2) \cdot Z_\gamma$$

$$y_\gamma = (1 - p_{\text{mis}}) \cdot Z_\gamma + (p_{\text{mis}}) \cdot Z_\delta + (p_{\text{mis}})/2) \cdot Z_\beta$$

$$y_\delta = (1 - p_{\text{mis}}) \cdot Z_\delta + (p_{\text{mis}}/2) \cdot Z_\gamma.$$

The formulation mentioned above provides a comprehensive structure for modeling the supply chain network, considering the probabilities of misclassification and reclassification, batch processing, recycling, remanufacturing, and repairing. Monte Carlo simulation can be used to solve this model by simulating the probabilities and flows of products through the network. The Monte Carlo method leverages repeated random sampling to simulate the behavior of a system under uncertain conditions. By generating numerous possible scenarios, we can estimate the performance metrics and outcomes of a complex process, where variability is inherent due to the randomness of the return process (in terms of both rate, which also affects the position in the product lifecycle as well as whether it is within warranty, and quality), and the potential for pre-inspection misclassifications of returned products. This technique allows decision-makers to anticipate the impact of these uncertainties on CL, transport planning, and inventory management, ultimately leading to more robust supply chain strategies.

Monte Carlo simulation is particularly useful for studying the effects of misclassification in categorizing returned products. For example, a probability of misclassification p_{mis} represents the likelihood that a retailer can incorrectly classify products as great (α), good (β), bad (γ), or beyond economical repair (BER, δ). By simulating these probabilities across various product categories and lifecycle stages (e.g., growth, maturity, and decline), the Monte Carlo method provides insights into how pre-inspection and misclassification affect downstream decisions, such as transport planning and load configurations. Through iterative simulation runs, a distribution of possible outcomes is generated, which helps quantify the risk and variability associated with misclassification errors and other uncertainties in the supply chain network.

9.4 Case Study

9.4.1 Real-Life Case Data from an Electronic Company

We abstract and expand upon the pure electronics remanufacturer case study and dataset located in the UK (see [18]) and consider the supply chain of one of their stock keeping units as the case study for our research, as follows: The company sells electronics products to its customers via a single retailer (using real demand data), and the customers can return their products to the retailer, according to a "time spent with customer" distribution (also using real data). If the returned product is within warranty (3 years), it is considered for repair; if not, it is considered for remanufacturing. For this purpose, we have used a dataset consisting of 9 years of monthly data, which we obfuscate by changing its scale and turning into weekly through a jittering process, alongside a "time spent with customer" distribution. Combined, these give us a time series of returns, which are each individually assigned a random condition (α, β, γ, or δ). The final input dataset is a product of a Monte Carlo simulation itself and consists of one time series of 9 years of weekly demand, four warranty return time series (one of each class), and four out-of-warranty time series (one of each class). Figure 9.3 shows the output of a Monte Carlo simulation when the warranty duration is equal to 3 years.

In this case study, we also consider the following assumptions. These include: (i) We have a 100% return rate for all products sold. This is due to the fact that there is a special contract between the manufacturer and the customer. Interested readers can refer to [18] for other cases where this assumption holds. (ii) Actual inspection can categorize the product's condition with 100% accuracy. (iii) Repaired products should be returned to customers via retailer within 2 weeks. (iv) We assume perfect substitution, whereby remanufactured products can be readily used to meet end demand. (v) Warranty returns that are beyond economical repair, and out-of-warranty returns over and above the demand for new/remanufactured products are sent to recycler. (vi) Finally, we consider the consolidation of products at the retailer (we run simulations without consolidation, and the impact of inspection without consolidation is negligent). For both warranty and out-of-warranty products, Fig. 9.4 represents the product flows between various entities.

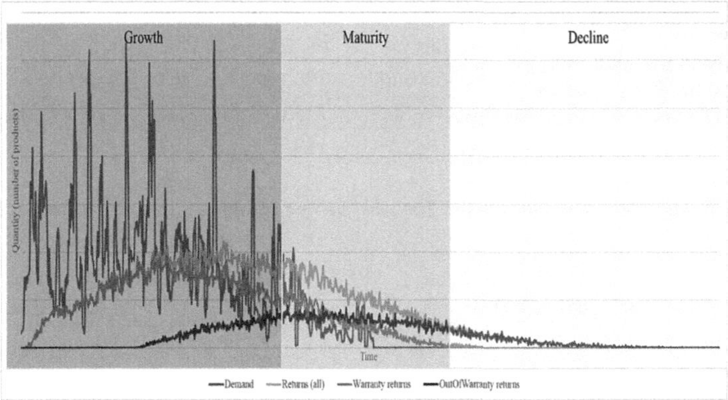

Fig. 9.3 Weekly demand, return, warranty return, out-of-warranty return across product lifecycle

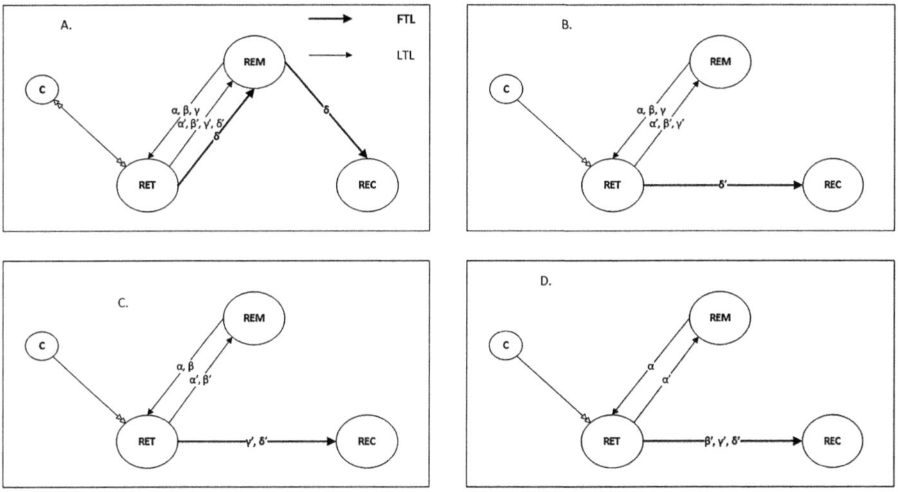

Fig. 9.4 A representation of circular logistics for the studied case study: (**a**) Warranty, (**b**) Out of warranty (growth), (**c**) Out of warranty (maturity), and (**d**) Out of warranty (decline)

9.4.2 Numerical Results

We now discuss the numerical results obtained from the case study. We note that the cases reported here were replicated with the absence of consolidation, because (a) had no impact on warranty returns (no consolidation means no FTL to remanufacturer even if inspected) and (b) were very similar to the no inspection results.

Table 9.4 Results for warranty repairs for various inspection regimes, total product, and truck trips across LTL and FTL types

Warranty	Number of LTL	Number of FTL	LTL truck trips	FTL truck trips
No inspection	133,005	9,967	4,683	298
$p_{mis} = 50\%$	123,038	19,934	4,385	596
$p_{mis} = 25\%$	119,737	23,578	4,314	704

Table 9.5 Results for warranty repairs for various inspection regimes, total product, and truck trips across LTL and FTL types

Warranty	RET to REM products traveled as LTL	RET to REM products traveled as FTL	RET to REC distance traveled (all FTL)
No inspection	51,110	0	0
$p_{mis} = 50\%$	41,560	9,550	0
$p_{mis} = 25\%$	40,654	11,267	0

9.4.2.1 Warranty Results

This section presents various KPIs for the warranty products in Tables 9.4 and 9.5.

The data in Table 9.4 summarizes the distribution of warranty product shipments and truck trips across different scenarios of missed inspections, indicated by p_{mis} values of 50% and 25%, as well as a baseline case without inspection. The total number of shipments is broken down by truckload type (LTL and FTL). In the "No inspection" case, there were 133,005 LTL and 9,967 FTL shipments, resulting in 4,683 LTL and 298 FTL trips. When p_{mis} is 50%, LTL shipments decrease slightly to 123,038, while FTL shipments double to 19,934, resulting in an increase in FTL trips to 596. This is the result of inspection allowing the δ stream of BER returns to be consolidated into FTL trips, which could not be done without inspection (or consolidation for that matter).

When p_{mis} is reduced to 25%, LTL shipments show a minor reduction to 119,737, which may be explained by the fact that the existence of FTL trips decreases the average number of products available for LTL shipments. FTL product shipments increase to 23,578 as fewer BER products are misclassified as bad, and similarly FTL trips increase at the expense of LTL trips. These results suggest that the reduction of inspection errors (lower p_{mis}) corresponds to an increase in the number of FTL trips, which indicate improvements in shipment consolidation practices as inspection accuracy improves.

We then zoom in to see what the impact of the different setups is to the transportation of products between stakeholders. When no inspection is conducted, all products from the retailer to the remanufacturer travel as LTL shipments, totaling 51,110 units as any inspection will have to take place with the remanufacturer. (There is no transportation from the retailer to the recycler as there is no understanding of which ones need to be recycled before they reach the remanufacturer for inspection.) Introducing a 50% misclassification rate reduces the LTL volume to 41,560 units, with 9,550 units shifting to FTL, as some products are classed as BER and sent to the remanufacturer via FTL. At a 25% misclassification rate,

LTL trips are further reduced to 40,654 units, while FTL trips increase to 11,267 units. This trend suggests that increasing inspection accuracy rates allows for better consolidation of shipments, potentially optimizing transportation costs and reducing the number of trips needed. There is no warranty shipment from retailer to recycler—all warranty repairs need to have a full inspection as a BER classification nulls the warranty.

9.4.2.2 Out-of-Warranty Results

We now move to out-of-warranty (remanufacturable) returns, as seen in Tables 9.6 and 9.7.

Similar to Tables 9.4 and 9.6 presents the impact of inspection regimes on out-of-warranty returns, showing the distribution of products and truck trips across LTL and FTL transportation modes. Without inspection, 36,272 products are transported via LTL and 11,820 via FTL, requiring 1,115 LTL and 150 FTL trips, respectively. When a 50% misclassification rate is introduced, the number of products in LTL drops significantly to 20,759, with a slight decrease in FTL to 9,646. This change reduces the LTL trips to 965 and doubles the FTL trips to 300, suggesting improved load consolidation with inspection. With a 25% misclassification rate, LTL product volume remains similar (20,457), while FTL slightly increases to 9,936, resulting in 959 LTL and 310 FTL trips. Overall, lowering misclassification rates (i) enables product to sidestep having to go from the retailer to the remanufacturer and (ii) shifts more products to FTL, increasing transportation efficiency and optimizing the logistics of out-of-warranty repairs.

Table 9.7 illustrates the distribution of out-of-warranty repair products across truckload types under different inspection regimes. With no inspection, all 15,480

Table 9.6 Results for out-of-warranty repairs for various inspection regimes, total product, and truck trips across LTL and FTL types

Outside warranty	Number of LTL	Number of FTL	LTL truck LTL trips	FTL truck trips
No inspection	36,272	11,820	1,115	150
$p_{mis} = 50\%$	20,759	9,646	965	300
$p_{mis} = 25\%$	20,457	9,936	959	310

Table 9.7 Results for out-of-warranty repairs for various inspection regimes, total product and truck trips across LTL and FTL types

Outside warranty	RET to REM products traveled as LTL	RET to REM products traveled as FTL	RET to REC distance traveled (All FTL)
No inspection	15,480	0	0
$p_{mis} = 50\%$	10,657	0	4,823
$p_{mis} = 25\%$	10,549	0	4,968

returned products travel as LTL shipments, with no products allocated as FTL (or recycling). At a 50% misclassification rate, the volume of products traveling as LTL decreases to 10,657, while 4,823 units are redirected to recycling, suggesting a more balanced distribution between remanufacturing and recycling. A further reduction in misclassification to 25% results in a similar LTL volume (10,549), with a slight increase in FTL shipments from the retailer to recycling to 4,968. These results indicate that inspection can help allocate returns more efficiently, reducing LTL shipments while promoting recycling when misclassification rates are lowered. There is no FTL transportation from the retailer to the remanufacturer as the products that need to be remanufactured (appropriate quality categories according to the position on the product lifecycle) are sent to the remanufacturer through LTL, and all other products are consolidated and sent to the recycler (FTL).

9.4.2.3 The Effect of Warranty Duration on Rate of Out-of-Warranty Returns

We present the details on the effect of warranty duration on rate of out-of-warranty returns in Fig. 9.5.

Figure 9.5 shows the comparison between demand and out-of-warranty returns over time for products with a 2-year warranty (left) and a 4-year warranty (right). As a reminder, the case analyzed above was for a warranty of 3 years. In both cases, demand is represented by the dark curve, while out-of-warranty returns are represented by the lighter curve. For the 2-year warranty, out-of-warranty returns begin to rise shortly after the initial demand surge and maintain a moderate level as demand tapers off. In contrast, with a 4-year warranty, out-of-warranty returns are delayed further, showing a smaller increase as demand declines. This translates into fewer and later out-of-warranty returns during the product's lifecycle, as well as more returned products being serviced under warranty.

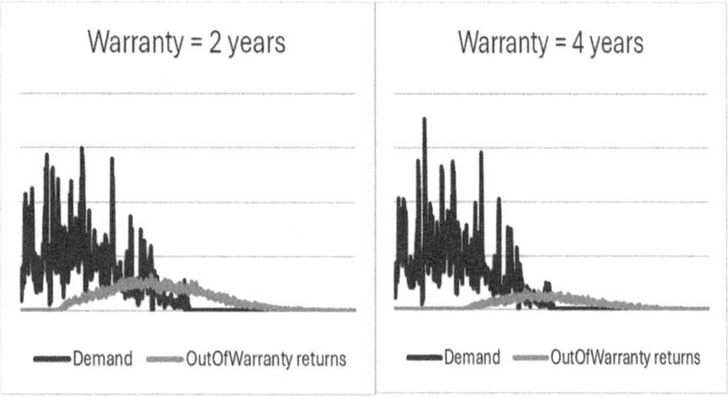

Fig. 9.5 Demand and out-of-warranty returns for 2 and 4 years

9.5 Discussion and Conclusion

We have investigated a (real) data-driven CL of a CE model consisting of customers, a retailer, a repair facility/remanufacturer, and a recycler, considering pre-inspection, consolidation, warranty, and a product's lifecycle. We reported a number of products and truck trips between the stakeholders with and without consolidation and pre-inspection, using forecasting to operationally define the position in a product's lifecycle (in particular with relation to remanufacturing operations) and quality characteristics.

We have the following insights derived from the real-life case study considered.

- As highlighted in the literature, the retailer is often the preferred collection channel, and pre-inspection at the retailer location makes huge improvements for the considered KPIs related to transport utilization.
- Consolidation and pre-inspection are both required to maximize the benefits of either. Consolidation without pre-inspection brings little benefits in terms of redirecting streams but does not allow for efficient truckload utilization.
- The impact of presorting accuracy on transportation loads exists but is not great on the repair (warranty) stream and is more prominent on the remanufacturing (outside warranty) stream.
- Forecasting is crucial in determining the position of a product in its lifecycle, but lifecycle stages need to be considered from a remanufacturing rather than a demand point of view.
- Warranty considerations heavily impact on the return mix (repair vs. remanufacturing), and its impact on these streams needs to be taken into account when setting it.

It is crucial to not forget that the current research does not consider the impact of above described scenarios on operations other than transportation, so the respective costs or benefits need to be appraised jointly with other impacts, such as the ones on repair and remanufacturing operations. As seen in the literature review, pre-inspection-enabled presorting has big impact on remanufacturing operations. As also seen in the literature review however, such works do not typically include transportation considerations and impacts. The transition to a CE has been shown to need systematic and systemic approaches and solutions so the need to consider these systems jointly is real, but beyond the scope of the current research.

This research highlights the role and potential of transportation, often considered a zero-value adding exercise in linear supply chain operations, to add value through well thought out and implemented CL. Our findings suggest that effective logistics and collaboration are critical components in overcoming the inherent challenges of transitioning to the CE, such as unpredictable material streams and cost efficiency in handling low-value materials.

Our investigation into pre-inspection, coordination, and lifecycle management reveals that these elements are often overlooked in conjunction with one another from a remanufacturer's viewpoint. Using real-life data from a leading global electronics company, we provided valuable insights into optimizing infrastructure and data-driven operational decision planning to return operations and enhancing the overall efficiency of the stylized closed-loop supply chain in question. The insights gained offer practical implications for industry stakeholders looking to adopt CE practices, in particular third-party logistics companies, suggesting that improvements in the collection channels and supply chain collaboration can significantly enhance CL operational performance.

Moving forward, this research opens several avenues for future exploration. Further studies could investigate the scalability of these strategies across different industries and geographical contexts by incorporating real geographical data and more realistic sets of stakeholders (e.g., multiple retailers, varied options for remanufacturing, etc.). Additionally, the impact of emerging technologies—such as blockchain and IoT on logistics in the CE—could be explored to better understand how these innovations might streamline processes and improve resource management. On one extreme, advanced sensors could potentially substitute the need for pre-inspection and supersede the need for forecasting [17].

Ultimately, as the global economy (one would hope) monotonically shifts toward sustainability at least partially in pursuit of UN's SDGs, embracing a circular approach not only presents a pathway to reducing environmental impacts and improving social sustainability, but also fosters economic resilience (often locally) in an ever-evolving market landscape.

References

1. Ghisellini, P., Cialani, C., Ulgiati, S.: A review on circular economy: the expected transition to a balanced interplay of environmental and economic systems. In: Journal of Cleaner Production, 114, pp. 11–32 (2016)
2. Geissdoerfer, M., Savaget, P., Bocken, N. M., Hultink, E. J.: The circular economy–A new sustainability paradigm?. Journal of Cleaner Production, **143**, 757–768 (2017)
3. Govindan, K., Soleimani, H.: A review of reverse logistics and closed-loop supply chains: A Journal of Cleaner Production focus. In: Journal of Cleaner Production, **142**, 371–384 (2017)
4. Farooque, M., Zhang, A., Liu, Y., Saeed, M. A.: Sustainable supply chain management through the lens of dynamic capabilities: A case of the circular economy. International Journal of Production Economics, **217**, 243–254 (2019)
5. Fahimnia, B., Sarkis, J., Davarzani, H.: Green supply chain management: A review and bibliometric analysis. International Journal of Production Economics, **162**, 101–114 (2015)
6. Stock, J. R., Boyer, S. L.: Developing a consensus definition of supply chain management: A qualitative study. International Journal of Physical Distribution & Logistics Management*, 39(8), pp. 690–711 (2009)
7. Kirchherr, J., Reike, D., Hekkert, M.: Conceptualizing the circular economy: An analysis of 114 definitions. In: Resources, Conservation and Recycling, **127**, 221–232 (2017)
8. Blomsma, F., Brennan, G.: The emergence of circular economy: A new framing around prolonging resource productivity. In: Journal of Industrial Ecology, **21**(3), 603–614 (2017)
9. Lacy, P., Rutqvist, J.: Waste to Wealth: The Circular Economy Advantage. Palgrave Macmillan UK (2016)
10. Savaskan, R.C., Bhattacharya, S., Van Wassenhove, L.N.: Closed-loop supply chain models with product remanufacturing. Management Science, **50**(2), 239–252 (2004)

11. Atasu, A., Toktay, L.B., Van Wassenhove, L.N.: How collection cost structure drives a manufacturer's reverse channel choice. Production and Operations Management, **22**(5), 1089–1102 (2013)
12. Choi, T.M., Li, Y., Xu, L.: Channel leadership, performance and coordination in closed-loop supply chains. International Journal of Production Economics, **146**(1), 371–380 (2013).
13. Schroeder, P., Anggraeni, K., Weber, U.: The relevance of circular economy practices to the sustainable development goals. Journal of Industrial Ecology, **23**(1), 77–95 (2019)
14. Potting, J., Hekkert, M. P., Worrell, E., Hanemaaijer, A.: Circular economy: measuring innovation in the product chain. Planbureau voor de Leefomgeving, (2544) (2017)
15. Kirchherr, J., Yang, N.H.N., Schulze-Spüntrup, F., Heerink, M.J., Hartley, K.: Conceptualizing the circular economy (revisited): an analysis of 221 definitions. Resources, Conservation and Recycling,
16. Sodhi, M. S., Tang, C.: Seeking and Exploiting Synergies Among the UN Sustainability Development Goals: Research Opportunities for Operations Management. Production and Operations Management, **33**(4), 962–978 (2024). https://doi.org/10.1177/10591478241235004
17. Goltsos, T. E., Ponte, B., Wang, S., Liu, Y., Naim, M. M., Syntetos, A. A.: The boomerang returns? Accounting for the impact of uncertainties on the dynamics of remanufacturing systems. International Journal of Production Research, **57**(23), 7361–7394 (2019). https://doi.org/10.1080/00207543.2019.1566665
18. Goltsos, T. E., Syntetos, A. A., van der Laan, E.: Forecasting for Remanufacturing: The Effects of Serialization. Journal of Operations Management, **65**(5), 447–467 (2019).
19. Van den Berg, M.R., Bakker, C.A.: A Product Design Framework for a Circular Economy. Product Lifetimes and the Environment, 365–379 (2015).
20. Atasu, A., Dumas, C., Van Wassenhove, L. N.: The circular business model. Harvard Business Review, **99**(4), 72–81 (2021)
21. Demir, E.: Value Creation through Green Vehicle Routing. In: Sustainable Freight Transport: Theory, Models, and Case Studies, pp. 63–78 (2018).
22. Guide, V. D. R.: Production planning and control for remanufacturing: industry practice and research needs. Journal of Operations Management, **18**(4), 467–483 (2000). https://doi.org/10.1016/s0272-6963(00)00034-6.
23. Ferguson, M., Guide, V. D., Koca, E., Souza, G. C.: The Value of Quality Grading in Remanufacturing. Production and Operations Management, **18**(3), 300–314 (2009). https://doi.org/10.1111/j.1937-5956.2009.01033.x
24. Aras, N., Boyaci, T., Verter, V.: The effect of categorizing returned products in remanufacturing. IIE Transactions, **36**(4), 319–331 (2004). https://doi.org/10.1080/07408170490279561
25. Azikopoulos, C., Tagaras, G.: On the attractiveness of sorting before disassembly in remanufacturing. IIE Transactions, **40**(3), 313–323 (2008). https://doi.org/10.1080/07408170701488078
26. Östlin, J., Sundin, E., Björkman, M.: Product life-cycle implications for remanufacturing strategies. Journal of Cleaner Production, **17**(11), 999–1009 (2009).
27. Van Wassenhove, L. N., Zikopoulos, C.: On the effect of quality overestimation in remanufacturing. International Journal of Production Research, **48**(18), 5263–5280 (2010). https://doi.org/10.1080/00207540903026384
28. Goh, T.N., Varaprasad, N.: A statistical methodology for the analysis of the life-cycle of reusable containers. IIE Transactions, **18**(1), 42–47 (1986).
29. Weetman, C.: A circular economy handbook for business and supply chains: Repair, remake, redesign, rethink. Kogan Page Publishers (2016).
30. Kaewpuang, R., Niyato, D., Tan, P.S., & Wang, P.: Cooperative management in full-truckload and less-than-truckload vehicle system. IEEE Transactions on Vehicular Technology, **66**(7), 5707–5722 (2016).

Chapter 10
Impact of Demand Forecasting on Environmental Performance of Coffee Supply Chain in Vietnam

Abdelghani Bekrar, Thi Thuy Hanh Nguyen, Thi-Muoi Le, and Mourad Abed

Abstract Customer demand variability is one of the most severe issues that lead to an imbalance between demand and supply. Therefore, demand forecasting is crucial in supply chain management because it provides potential trends, changes, and meaningful insights to eliminate uncertainty and stabilize the supply chain. Accurate demand forecasts have a significant impact on supply chain performance. In this chapter, we examine and compare the effect of demand forecasting techniques on environmental performance in the Vietnamese coffee supply chain context. We find that LSTM outperforms ARIMAX in terms of carbon and water footprint. As a result, advanced demand forecasting techniques are a valuable investment. However, managers and organizations need to consider several vital dimensions when implementing them. We provide some practical insights into the roles of demand forecasting techniques and issues that organizations may face.

Keywords Demand forecasting · Environmental performance · Coffee supply chain · Machine learning · Vietnam

---

A. Bekrar (✉) · M. Abed
LAMIH Laboratory, Université Polytechnique Hauts-de-France, Valenciennes, France
e-mail: Abdelghani.Bekrar@uphf.fr; Mourad.Abed@uphf.fr

T. T. H. Nguyen
University of Economics and Law and Vietnam National University, Ho Chi Minh City, Vietnam
e-mail: hanhntt@uel.edu.vn

T.-M. Le
CRISS Laboratory, Université Polytechnique Hauts-de-France, Valenciennes, France
e-mail: Thi-Muoi.Le@uphf.fr

B. Jarboui et al. (eds.), *Circular Economy Supply Chains*,
Computational Intelligence Methods and Applications,
https://doi.org/10.1007/978-981-96-7905-8_10

10.1 Introduction

Supply chain management (SCM) is related to managing a company's entire process flow of goods and business processes from upstream to downstream [17]. Demand forecasting is a primary SCM process that predicts future demands and customers' expectations [3]. Demand forecasting lays the foundation of supply chains [29] because SCM aims to provide the right product at the right place, time, and price to meet the expectations of the final customers [7]. Demand forecasting is critical in successfully implementing SCM strategies and activities. Based on the demand forecast information, the supply chain members make many proper operational, tactical, and strategic decisions in a variety of sectors, including demand planning, production planning, sales budgeting, new product launches, and promotion planning [1, 8, 23, 29].

Demand forecasting contributes to better the performance of supply chains. Forecasting provides potential trends, changes, and meaningful insights that assist businesses in developing better strategies through various forecasting activities (e.g., sales, financial, and demand forecasting). Accurate forecasting will minimize uncertainty and stabilize the supply chain [5]. It increases substantial financial savings, greater competitiveness, effective supply chain collaborations, and satisfies customers [29]. In contrast, incorrect predictions can lead to excessive expenditures on procurement and shipping, human resources, level of service, and inventories [1].

The agricultural supply chain has gained a lot of attention and concern because of its crucial role in the survival and development of people worldwide. However, there are many challenges and difficulties in managing the agricultural supply chain because of the many participants, the long production, harvesting processes, and complex storage. Moreover, climate change strongly threatens agricultural production [4], one of humanity's most serious environmental concerns [33]. Thus, building and developing sustainable supply chains are significant in an era of increasing global environmental concerns and saving scarce resources. Sustainable supply chain management (SSCM) has developed as a strategy for businesses to enhance long-term (environmental, social, and economic) results in their supply chains [16].

Many researchers have paid attention to coffee due to its role in national economies and global supply chains. Coffee is one of the world's most valued agricultural products and one of the most popular beverages [32]. The worldwide coffee market was worth 122.9 billion USD in 2022 and is predicted to reach 170.8 billion USD by 2030, growing at a CAGR of 4.2% from 2023 to 2030 (Global Information, 2024). Coffee is the primary income source of citizens in many developing countries, such as Vietnam, Indonesia, and Ethiopia [22, 26–28].

Moreover, demand for coffee products has increased yearly. Coffee demand is anticipated to triple by 2050 [11]. However, coffee production did not satisfy consumption for several years. Evidence shows that the coffee balance was negative in 2018/19, 2019/20, 2021/22, and 2022/23. The global coffee balance hit bottom

Table 10.1 Summary of the World Coffee Market [International Coffee Organization (ICO), 2023] Million 60-Kg Bags

	2018/19	2019/20	2020/21	2021/22	2022/23	2023/24
Production	169.8	168.4	170.8	168.0	168.2	178.0
Consumption	171.2	168.6	169.9	176.6	173.1	177.0
Balance	−1.3	−0.2	0.9	−8.6	−4.9	1.0
Growth rates, year on year						
Production	5.9%	−0.9%	1.4%	−1.7%	0.1%	5.8%
Consumption	3.3%	−1.5%	0.8%	4.0%	−2.0%	2.2%

in 2021/2022 during the Covid-19 pandemic. Thus, forecasting demand is vital for coffee supply chains (Table 10.1).

Nevertheless, there is a lack of research investigating the relationship between demand forecasting and sustainable supply chains. Previous research has paid more attention to the performance of forecasting models (e.g., prediction accuracy) than their impact on supply chain performance [10, 24, 31]. In addition, managers may consider investing in forecasting techniques a considerable expense. For example, researchers [19] examined forecasting practices over 20 years and discovered that the impact of forecasting methodologies on performance metrics (environmental indicators) was missing. Thus, this study assesses the influence of demand forecasting techniques on the sustainable supply chain to highlight and encourage their implementation.

Our research chose the coffee of Vietnam as a case study because Vietnam is the second largest coffee producer in the world (International Coffee Organization, 2023). According to [25], Vietnam's coffee supply chain was an average performance. One key reason is that they did not apply any forecasting techniques. Thus, employing forecasting techniques is vital to produce accurate demand forecasting for adequately planning and arranging associated inputs and resources, thereby boosting supply chain efficiency and effectiveness.

Therefore, we aim to provide empirical evidence of the effect of demand forecasting techniques on sustainable supply chains. Performance of sustainable supply chain will be measured by performance metrics by carbon emissions or water consumption. As a result, investing in forecasting techniques is beneficial in identifying market opportunities, strengthening channel relationships, increasing customer satisfaction, and improving the company's public image by being environmentally friendly.

Companies must efficiently include environmental concerns and sustainable practices in their daily operations, notably supply chain activities, in response to strict laws and growing customer awareness. However, there is a lack of research to assess the environmental performance of supply chains. As a result, our article evaluates environmental performance in the context of the coffee industry in Vietnam.

The rest of this chapter consists of five sections: The next section provides literature on demand forecasting techniques and environmental performance. Then,

Sect. 10.3 outlines the data collection method and empirical model. The main findings will be presented in Sect. 10.4. Section 10.5 discusses a company's challenges when implementing forecasting techniques. The final section will discuss the conclusion.

10.2 Literature Review

Various predictive methods and techniques have been developed to anticipate future demand. According to prior researchers [14], forecasting methods can be divided into traditional and machine learning (ML). Each forecasting method has advantages and limitations compared to others [2]. A recent study [1] emphasized that no one approach can overcome all forecasting issues and provide the best forecasting results of all the models.

Our study focuses on the impact of demand forecasting techniques (ARIMAX and LSTM) on carbon and water footprint in the context of coffee in Vietnam (as shown in Table 10.2). We chose the autoregressive integrated moving average with exogenous variables (ARIMAX) to represent the traditional approach and the long short-term memory (LSTM) to represent the ML method.

ARIMAX models are among the most common oldest linear models used for time series forecasting [13]. In contrast, LSTM is a recently famous neural network due to its ability to cope with long-term dependency [9]. Nonetheless, no study has been conducted to compare the influence of each model on environmental performance to be the best of authors' knowledge. Thus, we would like to fill in the gaps.

Table 10.2 illustrates how ARIMAX and LSTM differ in complex data, training period, and nonlinear trends. As can be seen, each model has advantages and disadvantages. ARIMAX takes less time to train than LSTM. In contrast, the LSTM model performs well with complex data and nonlinear trends.

According to [6], environmental performance is one of three critical elements of sustainability. It is interpreted as the capability to fulfill current demands without

Table 10.2 Comparison of forecasting model characteristics [12]

Forecasting model	Model group	Characteristics	Complex data	Training period	Nonlinear trend
ARIMAX	Regression	This model was developed from the ARIMA model but also considered exogenous factors	++	++	+
LSTM	Neural network	This model is based on the concept of a memory cell	++++	+	+++

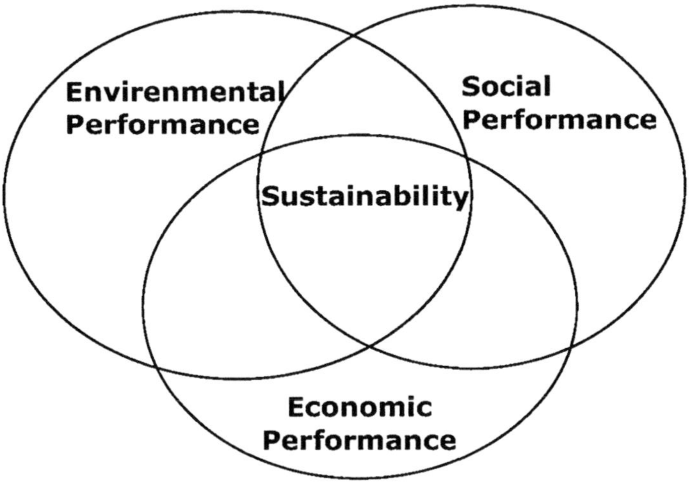

Fig. 10.1 Sustainability: the triple bottom line [6]

compromising the ability of future generations to meet their needs (as shown in Fig. 10.1). Besides, [15] clarified that the environmental aspect of sustainability is ecologically friendly production (reuse, reverse, and recycle) in the long term without polluting the environment.

In literature, various indicators can be used to measure environmental performance. Carbon footprint and water footprint are commonly used to assess the ecological impacts of any processes or products on greenhouse gas emissions and water consumption [30]. As a result, in this chapter, we will use carbon and water footprints to assess environmental performance. In the following section, we will discuss each indicator in detail.

Carbon Footprint

Most prior research measured environmental performance by analyzing primary data [20]. For example, [20] collected data from peer-reviewed literature, primary data from coffee producers, and interviews with coffee farmers. As a result, assessing the impact of forecasting techniques on environmental performance is critical for promoting the adoption of technological innovations in coffee supply chains. This research will measure environmental performance by assessing carbon (Fig. 10.2).

The carbon footprint is recognized as a valuable indicator of greenhouse gas emissions [33]. Understanding greenhouse gas emissions from the coffee supply chains is critical for evaluating options for climate change mitigation within the sector [18]. According to authors [20], the carbon footprint of Vietnamese coffee is 16040 tons of CO_{2e}/ton of green coffee.

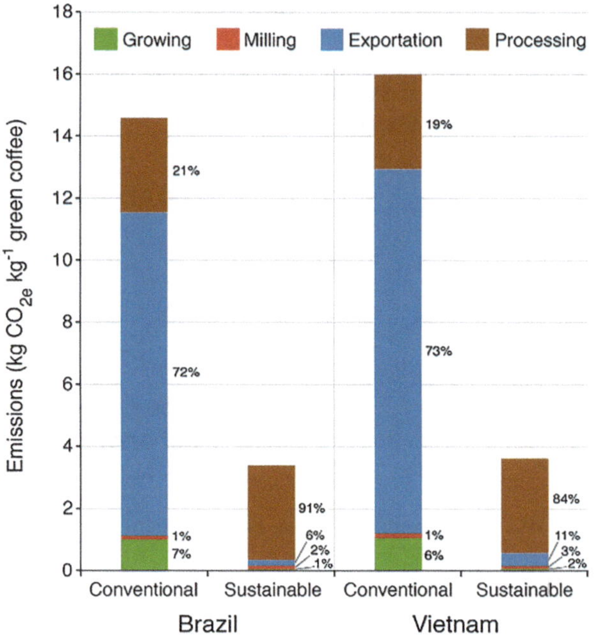

Fig. 10.2 Coffee carbon footprint of Vietnamese coffee [20]

Water Footprint

Due to the impact of climate change, the water stress situation in Vietnam has become more severe due to the high water consumption of coffee planting and production. Furthermore, Vietnam's location downstream of international river basins makes it more vulnerable to the water policies of upstream countries such as China and Lao. Water pollution from wet processing is one of the significant issues in coffee production [International Coffee Organization (ICO), 2022]. According to [21], the water footprint per ton of green coffee in Vietnam is 12,829 m3/ton.

10.3 Methodology

This chapter employs a quantitative method to investigate the influence of forecasting techniques on environmental performance. Figure 10.3 depicts the steps of our research.

First, we collect data from The International Coffee Organization (ICO), an intergovernmental coffee organization (https://ico.org/resources/trade-statistics-tables/).

The data is then divided into training and testing [10, 12]. After tuning hyperparameters, the best configuration of LSTM will be chosen. ARIMAX and LSTM

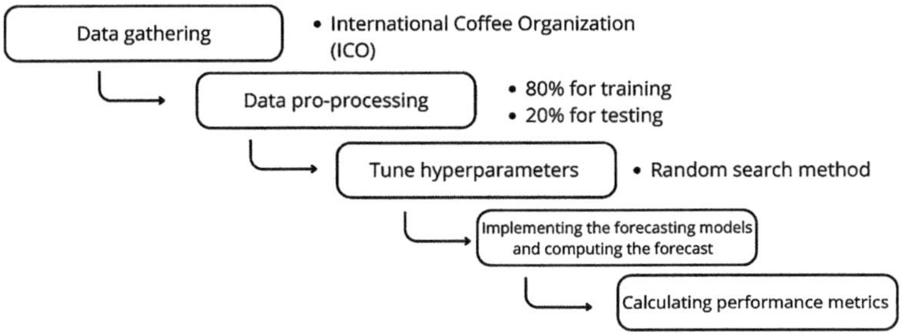

Fig. 10.3 Steps for calculating environmental performance

will then forecast coffee demand. Finally, the following two environmental indicators will be computed as follows:

$$c_t = y_t \times 16040 \tag{10.1}$$

$$w_t = y_t \times 12829 \tag{10.2}$$

$$\widehat{c_t} = \widehat{y_t} \times 16040 \tag{10.3}$$

$$\widehat{w_t} = \widehat{y_t} \times 12829 \tag{10.4}$$

$$Diff\_CFP = \frac{\sum_t^n |c_t - \widehat{c_t}|}{n} \tag{10.5}$$

$$Diff\_WFP = \frac{\sum_t^n |w_t - \widehat{w_t}|}{n}, \tag{10.6}$$

where:

- $Diff\_CFP$ is the difference in the carbon footprint.
- $Diff\_WFP$ is the difference in the water footprint.
- $\widehat{y_t}$ is the forecast demand at time t (ton green beans).
- y_t is the real demand at time t (ton green bean).
- c_t is the carbon footprint at time t.
- $\widehat{c_t}$ is the carbon footprint at time t.
- w_t is the water footprint at time t.
- $\widehat{w_t}$ is the water footprint at time t.

10.4 Experimental Results

Table 10.3 compares the carbon footprint and water footprint of ARIMAX and LSTM. In both indicators, LSTM has smaller difference values than ARIMAX. Thus, LSTM outperforms ARIMAX in terms of environmental performance. It also implies that different forecasting techniques produce different results. More

Table 10.3 Environment performance based on different forecasting techniques

ARIMAX	LSTM	Difference ARIMAX-LSTM
The difference in carbon footprint between the predicted value and the actual value (ton CO_{2e}/ton green coffee)		
12032399.12	10501095.44	1531303.68
The difference in water footprint between the predicted value and the actual value (m^3/ton green bean)		
9623668.849	8398912.303	1224756.546

advanced forecasting techniques improve environmental performance by lowering carbon and water footprints.

10.5 Implications and Potential Issues

Based on the obtained results of demand forecasting using both LSTM and ARMIMAX models, some insights and challenges for implementing forecasting techniques for sustainable supply chains are given in the following section.

10.5.1 Insights into Demand Forecasting Applications

As we can see, a good forecasting of the supply chain demand can improve supply chain sustainability effects. However, other insights could be explored in future studies, some of which are listed below.

Exploiting Advantages of Machine Learning
Forecasting is critical activity in demand planning and supply chain management. Big data technology can help organizations analyze large amounts of data intelligently and quickly. As a result, businesses would gain a comprehensive understanding of customer trends by extracting valuable patterns and information. Big data technology has developed and produced many advanced forecasting techniques. They help managers conduct in-depth market analyses, allowing them to plan production, financing, labor, and marketing better. New forecasting techniques are becoming more robust and accurate than older ones. As a result, businesses should not disregard forecasting techniques.

Furthermore, better forecasting techniques will aid in optimizing the performance of businesses and supply chains. They improve not only operational and financial performance but also environmental performance. Moreover, they can promote environmental sustainability by allowing the company to reduce its energy and water needs. This contribution is becoming increasingly important as the population grows, demand rises, and climate changes. As a result, forecasting with big data technology has the potential to improve efficiency and quality in the coffee supply chain from upstream to downstream, driving it toward sustainable growth [15].

Reducing Carbon Emission

According to our findings, using advanced forecasting techniques can result in a lower carbon footprint, thereby mitigating the effects of global climate change. Coffee is the primary income of smallholder farmers in developing countries. However, climate change is putting their lives at risk. Furthermore, companies are part of the solution to climate change. Companies should make reducing the impact of climate change a long-term goal because they would not survive if there were no inputs or materials for production.

Moreover, consumers' purchasing decisions are shifting toward environmentally friendly products. Companies will not sell products that do not meet the needs and concerns of their customers. Thus, investing in advanced forecasting techniques benefits companies economically and helps them develop an eco-friendly image.

Sustainable Water Resources Management

Water is critical for socio-economic development, healthy ecosystems, and human survival. As a result, reducing water consumption is now a vital management consideration for all organizations. Water conservation can benefit businesses, including lower costs, reduced water scarcity, and a cleaner environment. Furthermore, the company can improve its public image, attract new loyal customers, and increase its competitiveness.

Using advanced forecasting techniques, managers can accurately predict future water consumption and draw up appropriate plans to use water resources proactively and allocate water efficiently.

10.5.2 Challenges and Potential Concerns for Implementing New Demand Forecasting Techniques

To improve the sustainability impact of companies' supply chains by improving demand forecasting, there is no unique, perfect model to adopt. The implementation of new models, particularly those based on machine learning, requires certain requirements to be taken into account.

Resistance to Change

Although forecasting is critical in organizations, many companies face challenges when implementing them due to several factors. These reasons, however, are raised by both managers and employees.

Managers' main reasons include a lack of confidence, financial support, and self-interest.

First, some senior managers recognize the benefits of advanced technology; however, because technology evolves rapidly, some are unsure how to use it. Also, senior managers are often good at management but lack engineering knowledge. As a result, they lack confidence in the value of new technology change (for example, whether investing in complicated forecasting techniques is worth the expense or not). As a result, they ignore or resist change.

Second, several leaders understand the advantages of technological applications. They see the benefits that the change will bring. They can build a better future. However, some managers cannot make investment decisions because they may impact shareholder profits. Some investors, particularly risk-averse investors, are unwilling to invest in new technological practices because they are more concerned with earnings per share. As a result, managers must persuade and demonstrate the return on investment that the company will receive.

Third, some managers may feel lost control or powerless. Thanks to technological implementations, companies can make better decisions based on forecasting data; thus, some lower-level managers or workers can be replaced. To solve this issue, managers must develop the right mindset for themselves and those around them. They are using advanced forecasting techniques as an opportunity to strengthen the business by aligning operations with strategy, taking on new professional challenges and risks, and advancing their careers.

Employees resist learning and using new technological techniques due to a lack of competence and a communication strategy.

First, employees must learn how to use and develop the necessary skills. As a result, they fear they will not be able to perform as well as under previous systems. In consequence, they risk losing their jobs, as some positions may be downsized.

Second, employees do not understand the roles, policies, and strategies when implementing new forecasting systems. As a result, poor communication during the transition leads to dissatisfied and disengaged employees. Thus, to address this issue, businesses must engage in open dialogue to promote transparency and ease the transition from old to new technology. Companies, for example, communicate the benefits of new technology and encourage questions and feedback so that managers can understand what their employees think and deal with any problems.

Impact of External Events

Although advanced forecasting techniques aid in accurately forecasting customer trends, some unexpected events may influence consumer behavior decisions. For example, explore product issues that may harm one's health. Furthermore, some circumstances, such as a pandemic, war, or foreign sanctions, are beyond the company's control.

Adaption to the Company

There are numerous forecasting techniques available. However, not all methods are appropriate for every business. Thus, before implementing new forecasting techniques, managers should be aware of the advantages and disadvantages of each method.

Furthermore, managers must understand the organization's situation, requirements, and goals. For example, investing in the most advanced tools will be ineffective if the company has small data or a small budget. However, the company can combine and use different forecasting techniques before making plans or decisions.

10.6 Conclusions

Demand forecasting plays an essential role in supply chain management. Hence, numerous forecasting techniques have been developed to aid demand forecasting and planning operations. This chapter provides empirical evidence regarding the impact of forecasting techniques on environmental performance.

We compare the impact of two well-known techniques, ARIMAX and LSTM. We discovered that LSTM, as a machine learning technique, reduces carbon and water footprints better than ARIMAX. As a result, investing in advanced forecasting techniques is worthwhile in the context of Vietnam's coffee supply chain. We also discuss demand forecasting techniques, and the challenges businesses may encounter when applying new forecasting tools.

Although environmental performance is estimated based on demand, companies can benefit from advanced forecasting techniques, lower carbon emissions, and water consumption. We also believe that the identified issues can assist supply chain managers in various supply chains.

References

1. Abolghasemi, M., Beh, E., Tarr, G., Gerlach, R. (2020). Demand forecasting in supply chain: The impact of demand volatility in the presence of promotion. Computers and Industrial Engineering, 142(July 2019), 106380. https://doi.org/10.1016/j.cie.2020.106380
2. Aburto, L., Weber, R. (2007). Improved supply chain management based on hybrid demand forecasts. Applied Soft Computing Journal, 7(1), 136–144. https://doi.org/10.1016/j.asoc.2005.06.001.
3. Benkachcha, S., Benhra, J., El Hassani, H. (2014). Demand forecasting in supply chain: Comparing multiple linear regression and artificial neural networks approaches. International Review on Modelling and Simulations, 7(2), 279–286.https://doi.org/10.15866/iremos.v7i2.641.
4. Birkenberg, A and Birner, R (2018). The world's first carbon neutral coffee: Lessons on certification and innovation from a pioneer case in Costa Rica. Journal of Cleaner Production, 189, 485–501. https://doi.org/10.1016/j.jclepro.2018.03.226.
5. Bhadouria, S., Jayant, A. (2017). Development of ANN Models for Demand Forecasting. American Journal of Engineering Research (AJER), 12, 2012–2017.
6. Carter, Craig R. and Rogers, Dale S. (2008). A framework of sustainable supply chain management: moving toward new theory, International Journal of Physical Distribution and Logistics Management 5(38), 360–387.
7. College, D. S., Helms, M. M., Chapman, S. (2000). supply chain management Supply chain forecasting. Business Process Management Journal, 6(5), 392–407.
8. Danese, P., Kalchschmidt, M. (2011). The role of the forecasting process in improving forecast accuracy and operational performance. International Journal of Production Economics, 131(1), 204–214. https://doi.org/10.1016/j.ijpe.2010.09.006.

9. Dave, E., Leonardo, A., Jeanice, M., Hanafiah, N. (2021). ScienceDirect Forecasting Indonesia Exports using a Hybrid Model ARIMA- LSTM. Procedia Computer Science, 179(2020), 480–487. https://doi.org/10.1016/j.procs.2021.01.031.
10. Feizabadi, J. (2020). Machine learning demand forecasting and supply chain performance. International Journal of Logistics Research and Applications, 0(0), 1–24. https://doi.org/10.1080/13675567.2020.1803246.
11. Grant, T. (2021). What Is The Carbon Footprint Of a Cup Of Coffee? MTPAK, 1. https://mtpak.coffee/coffee-roasters-reduce-carbon-footprint/.
12. Kantasa-ard, A., Nouiri, M., Bekrar, A., Ait el cadi, A., Sallez, Y. (2021). Machine learning for demand forecasting in the physical internet: a case study of agricultural products in Thailand. International Journal of Production Research, 59(24), 7491–7515. https://doi.org/10.1080/00207543.2020.1844332.
13. Khashei, M., Bijari, M. (2011). A novel hybridization of artificial neural networks and ARIMA models for time series forecasting. Applied Soft Computing Journal, 11(2), 2664–2675. https://doi.org/10.1016/j.asoc.2010.10.015.
14. Kilimci, Z. H., Akyuz, A. O., Uysal, M., Akyokus, S., Uysal, M. O., Atak Bulbul, B., Ekmis, M. A., Silva, T. C. (2019). An improved demand forecasting model using deep learning approach and proposed decision integration strategy for supply chain. Complexity, 2019. https://doi.org/10.1155/2019/9067367.
15. Kittichotsatsawat, Y., Jangkrajarng, V., Tippayawong, K. Y. (2021). Enhancing coffee supply chain towards sustainable growth with big data and modern agricultural technologies. Sustainability (Switzerland), 13(8), 1–20. https://doi.org/10.3390/su13084593.
16. Koberg, E., Longoni, A. (2019). A systematic review of sustainable supply chain management in global supply chains. Journal of Cleaner Production, 207, 1084–1098. https://doi.org/10.1016/j.jclepro.2018.10.033.
17. Kusrini, E., Caneca, V. I., Helia, V. N., Miranda, S. (2019). Supply Chain Performance Measurement Using Supply Chain Operation Reference (SCOR) 12. Model: A Case Study in A A Leather SME in Indonesia. IOP Conference Series: Materials Science and Engineering, 697(1), 0–10. https://doi.org/10.1088/1757-899X/697/1/012023.
18. Maina, J. J., Mutwiwa, U. N., Kituu, G. M., Githiru, M. (2015). Evaluation of Greenhouse Gas Emissions along the Small-Holder Coffee Supply Chain in Kenya. Open Access Journal of Sustainable Research in Engineering, 2(4), 111–120.
19. Mccarthy, T. M., Davis, D. F., Golicic, S. L., Mentzer, J. T. (2006). The evolution of sales forecasting management: A 20-year longitudinal study of forecasting practices. Journal of Forecasting, 25(5), 303–324. https://doi.org/10.1002/for.989.
20. Nab, C. and Maslin, M. (2020). Life cycle assessment synthesis of the carbon footprint of Arabica coffee: Case study of Brazil and Vietnam conventional and sustainable coffee production and export to the United Kingdom. Geo: Geography and Environment, 7(2), 1–19. https://doi.org/10.1002/geo2.96.
21. Ngo, T. T., Le, N. T., Hoang, T. M., Luong, D. H. (2018). Water Scarcity in Vietnam: a Point of View on Virtual Water Perspective. Water Resources Management, 32(11), 3579–3593. https://doi.org/10.1007/s11269-018-2007-4
22. Nguyen, G. N. T and Sarker, T. (2018). Sustainable coffee supply chain management: a case study in Buon Me Thuot City, Daklak, Vietnam. International Journal of Corporate Social Responsibility, 3(1), 1–17. https://doi.org/10.1186/s40991-017-0024-x
23. Nguyen, T. T. H. (2023). Applications of Artificial Intelligence for Demand Forecasting. Operations and Supply Chain Management, 16(4), 424–434. https://doi.org/10.31387/oscm0550401
24. Nguyen, T. T. H., Bekrar, A., Abed, M., Le, T. (2021). Supply Chain Performance Measurement using SCOR Model: a Case Study of the Coffee Supply Chain in Vietnam. 2021 1st International Conference On Cyber Management And Engineering (CyMaEn), 1–7. https://doi.org/10.1109/CyMaEn50288.2021.9497309

25. Nguyen, T. T. H., Bekrar, A., Le, T. M., Abed, M. (2021). The Effect of Machine Learning Demand Forecasting on Supply Chain Performance—The Case Study of Coffee in Vietnam. In H. M. Le Thi, H.A., Pham Dinh, T., Le (Ed.), Lecture Notes in Networks and Systems, vol 363. (pp. 247–259). Springer, Cham. https://doi.org/10.1007/978-3-030-92666-3
26. Nguyen, T. V., Nguyen, N. C., Bosch, O. J. H. (2015). Contribution of the systems thinking approach to reduce production cost and improve the quality of Vietnamese coffee. International Journal of Markets and Business Systems, 1(1), 53–67. https://doi.org/10.1504/ijmabs.2015.070272
27. Nguyen, T. Van, Nguyen, N. C., Bosch, O. J. H. (2017). Identifying key success factors in supply chain management for increasing the competitive advantages of Vietnamese coffee. An International Business Journal, 27(5), 438–461.
28. Palomino, E. R., Meza, S. H., Raffo, D. R. M. F. S., Lo, W.-S. (2017). Organic Coffee Supply Chain Management in the San Martin Region of Peru. International Journal of Innovation, Management and Technology, 9–16. https://doi.org/10.18178/ijimt.2017.8.1.694
29. Perera, H. N., Hurley, J., Fahimnia, B., Reisi, M. (2019). The human factor in supply chain forecasting: A systematic review. European Journal of Operational Research, 274(2), 574–600. https://doi.org/10.1016/j.ejor.2018.10.028
30. Ratchawat, T., Panyatona, S., Nopchinwong, P., Chidthaisong, A., Chiarakorn, S. (2020). Carbon and water footprint of Robusta coffee through its production chains in Thailand. Environment, Development and Sustainability, 22(3), 2415–2429. https://doi.org/10.1007/s10668-018-0299-4
31. Soy Temür, A., Akgün, M., Temür, G. (2019). Predicting housing sales in Turkey using ARIMA, LSTM and hybrid models. Journal of Business Economics and Management, 20(5), 920–938. https://doi.org/10.3846/jbem.2019.10190
32. Valencia-Payan, C., Grass-Ramirez, J. F., Ramirez-Gonzalez, G., Corrales, J. C. (2022). A Smart Contract for Coffee Transport and Storage With Data Validation. IEEE Access, 10, 37857–37869. https://doi.org/10.1109/ACCESS.2022.3165087
33. Williams, M. (2012). Tackling climate change: What is the impact on air pollution? Carbon Management, 3(5), 511–519. https://doi.org/10.4155/cmt.12.49

Chapter 11
Exploring Global Food Security Index: A Case Study of Indonesia

Muhammad Younus, Achmad Nurmandi, Dyah Mutiarin, Andi Luhur Prianto, Halimah Abdul Manaf, Danang Kurniawan, Shafira Mourina Yasmine, Rendra Agusta, and Triwiyanti Triwiyanti

Abstract This research study analyzes the Global Food Security Index (GFSI) for Indonesia, a developing country with a fast-expanding economic landscape. The GFSI is a comprehensive methodology that assesses a country's food security readiness, resilience, and response capabilities. The study's approach includes a detailed assessment of GFSI data, as well as related literature and policy papers. The analysis focuses on major GFSI dimensions such as affordability, availability, quality and safety, and sustainability and adaptation. Furthermore, the study considers Indonesian socioeconomic and political settings to better understand the elements driving their food security performance. Based on the exploratory analysis, this study presents recommendations for Indonesia to improve its food security posture. Among them are enhancing legal frameworks, investing in technical infrastructure, encouraging capacity-building efforts, and increasing regional and international

M. Younus (✉)
Department of Government Affairs and Administration, University of Muhammadiyah Yogyakarta, Yogyakarta, Indonesia
e-mail: nurmandi_achmad@umy.ac.id; dyahmutiarin@umy.ac.id

A. Nurmandi · D. Mutiarin
Department of Government Affairs and Administration, University of Muhammadiyah Yogyakarta, Yogyakarta, Indonesia

Department of Product Research and Software Development, TPL Logistics Pvt Ltd, Karachi, Pakistan

A. L. Prianto
Department of Government Studies, Universitas Muhammadiyah Makassar, Makassar, Indonesia
e-mail: luhur@unismuh.ac.id

H. A. Manaf
UUM College of Law, Government, International Studies Universiti Utara Malaysia, Kedah, Malaysia
e-mail: imah@uum.edu.my

D. Kurniawan, S. M. Yasmine, R. Agusta, T. Triwiyanti
PT. Sinergi Visi Utama, Yogyakarta, Indonesia

B. Jarboui et al. (eds.), *Circular Economy Supply Chains*,
Computational Intelligence Methods and Applications,
https://doi.org/10.1007/978-981-96-7905-8_11

cooperation. By implementing these measures, Indonesia can better protect its people from food crises and contribute to more secure food security. The study attempts to uncover the strengths and weaknesses of Indonesia's food security posture and provide insights into potential areas for improvement by studying the GFSI scores and underlying indicators.

Keywords Global Food Security Index · Food Crises · Food Security · Indonesia

11.1 Introduction

Food security is a complex and crucial matter that represents one of the most urgent global concerns in contemporary times. The concept of food security involves not only the presence of an adequate food supply [7] but also factors such as the ease of obtaining food, its price, and its overall quality. With the projected worldwide population growth from 7.8 billion in 2021 to roughly 10 billion by 2050, as estimated by the United Nations, the need for food is intensifying. This further complicates the already delicate task of guaranteeing food security [15].

This research study explores the complex issue of global food security [5], with a particular emphasis on Indonesia, a nation characterized by a dynamic and diversified environment, in terms of both geography and culture. Indonesia's dedication to ensuring food security holds significant importance in its economic and social progress, considering its position as the fourth most populous nation globally. The Global Food Security Index (GFSI) [14], an innovative and comprehensive instrument developed by The Economist Intelligence Unit (EIU), provides a quantitative assessment of food security in countries around the world. The foundation of this ranking is established upon three fundamental principles: price, availability, quality, and safety. The framework consists of multiple indicators within each pillar, allowing for a comprehensive examination of a country's food security status. Examining the Global Food Safety Initiative (GFSI) [13] and its ramifications for Indonesia presents a useful occasion to evaluate the country's capacities and deficiencies in guaranteeing food security, as well as to pinpoint potential avenues for action.

Food security is not solely concerned with addressing hunger; rather, it is an essential requirement for establishing a stable, thriving, and well-being-oriented society. Food security at the global level is intricately connected to other sustainable development goals (SDGs), such as zero hunger, improved health, economic advancement, and decreased disparities. Comprehending and effectively tackling the issue of food security is crucial in attaining a sustainable and just global society.

Food security in Indonesia is of utmost importance, given the substantial magnitude of its people. Indonesia, renowned for its extensive cultural heterogeneity, accommodates a population of 270 million individuals, each characterized by distinct culinary inclinations and dietary needs. The presence of diverse food sources presents both obstacles and opportunities in the context of food security [18]. The complicated nature of guaranteeing the accessibility of culturally appropriate and nutritionally sufficient food options adds to the complexity of this undertaking.

Additionally, the expansive geography of Indonesia, encompassing both highly crowded urban centers and isolated rural regions, gives rise to inequalities in terms of food accessibility and availability. Gaining a comprehensive understanding of how the nation strategically manages these intricate geographical complications is crucial for the development and implementation of good policy planning. Moreover, the ramifications of climate change and environmental degradation (Randell & Southard, 2022) [22] provide a substantial peril to agricultural output and food supply chains in Indonesia, hence rendering food security an urgent matter.

11.1.1 Research Objectives and Scope

The primary objective of this research is to investigate the Global Food Security Index and its relevance within the Indonesian environment. The main goals that we aim to achieve are:

- This study aims to conduct a comprehensive examination of Indonesia's performance in the Global Food Security Index, taking into account its distinct geographical, cultural, and economic attributes.
- The objective of this study is to assess the strengths and weaknesses of Indonesia's food security situation using the Global Food Security Index (GFSI) indicators. This analysis aims to provide a comprehensive understanding of the country's food security landscape, which can then influence evidence-based policy suggestions.
- The objective of this study is to analyze the potential consequences of climate change and environmental issues on the food security situation in Indonesia, as well as to evaluate the country's ability to withstand and adapt to these risks.
- The objective is to provide policy recommendations and strategic insights that will effectively enhance food security in Indonesia. This, in turn, will contribute to the broader global objective of achieving zero hunger and sustainable development.

11.2 Literature Review

According to the existing literature, social capital is believed to have a significant impact on the enhancement of household food security. Previous research suggests that social capital plays a crucial role in enhancing the resilience of households by providing them with valuable social and psychological support [27]. The presence of social cohesiveness and mutual trust within a community has been found to enhance individuals' access to various resources, external assets, job possibilities, and the transmission of information. This is mostly achieved through the facilitation of assistance-seeking behaviors among community members, who are more likely to seek help from their neighbors and relatives during hard situations [1]. Numerous scholarly investigations have been undertaken to validate the influence of social capital on food security [24]. Several studies have been undertaken in different locations, such as the United States and Nigeria. For instance, Haysom and Tawodzera [11] did a study in the United States, while Vaitla et al. [25] focused their research specifically on the state of Texas. Additionally, Mekonnen and Gerber [17] carried out a study in Ethiopia. The results suggest that social capital plays a substantial role in influencing the level of food security [10]. It discovered that social capital exerts a favorable and noteworthy influence on the level of food security within households in Texas. The involvement of those with limited financial resources in community engagement is notable, and social activities demonstrate a greater inclination toward providing assistance during periods of adversity within the community. According to Devereux et al. [8], there exists a correlation between the social capital of American households and their level of food security. Several studies undertaken in Indonesia have yielded comparable results. Nevertheless, the scope of the studies is restricted to the selection of homes within specific regions of Indonesia. Numerous scholarly investigations have focused only on certain provinces or districts, exemplified by Haysom and Tawodzera [11] who conducted a comprehensive examination of the Maluku Islands, who conducted an in-depth analysis of Pidie Regency. Prior studies have only focused on certain villages, namely Martianto, Alfiasari, and Dharmawan, who conducted research in the Tanah Sareal subdistrict of Bogor, as well as Prayitno, Baiq Maulida, and Nugraha and Nawrotzki et al. [18], who explored the Ngadirejo subdistrict in Malang. The present circumstances have had an effect on the applicability of previous research findings in informing the formulation of national policies pertaining to matters of food security [19]. According to the Department of Agriculture and Food Security (DPKP), in 2019, several instances of crop failures were observed primarily in Gunungkidul, which were attributed to drought conditions. The quantity of rice plants that do not undergo successful harvesting has a significant impact on food security, particularly with regard to food availability. This particular component holds significant importance for two primary reasons. First, insufficiency in food supply can result in inadequate nutrition, leading to deficits in essential nutrients. Second, an excessive consumption of undernourished food has the potential to induce various health issues, including obesity, diabetes, and hypertension (Figs. 11.1, 11.2, and Table 11.1).

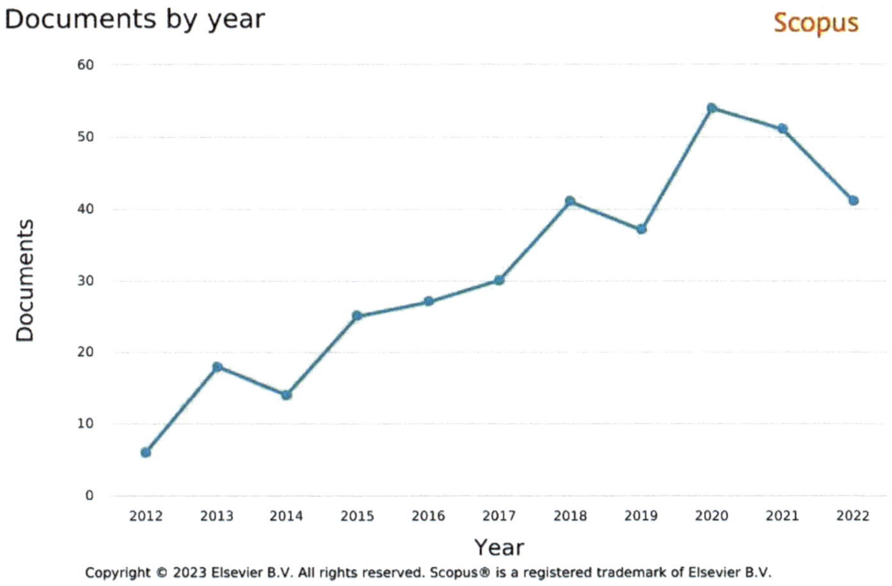

Fig. 11.1 Publications by years—Source Scopus Database

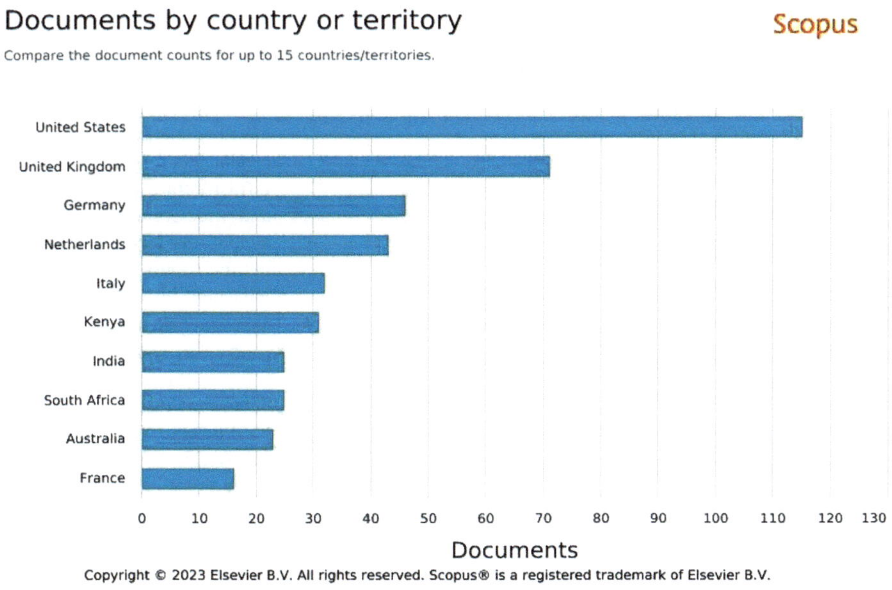

Fig. 11.2 Publications by country—Source Scopus Database

Table 11.1 Taxonomy of research literature

Paper title	Authors	Year	Main findings
Indonesian food security policy	M. Limenta, S. Candra [16]	2017	Indonesian policies have focused on promoting self-sufficiency in food production to achieve food security. However, this policy has not effectively ensured availability and access to nutritious food. More effective and efficient solutions are needed
The law no. 18/2012 governing food security in Indonesia	I. Rafai [21]	2014	Food is a basic human right and should be available at affordable prices. Law No. 18/2012 was established to protect producers and consumers, ensuring food does not conflict with religion, belief, and culture
Food security in Indonesia: a continued reliance on foreign markets	Mervyn Piesse [20]	2016	Indonesia has failed to achieve agricultural development due to restrictive import policies. Increasing interisland connectivity while supplementing food supplies with imports can improve food security
Food security in Indonesia	H. S. Dillon [9]	2004	Since the 1970s, Indonesia has promoted food production through infrastructure, buffer stocks, and price control. Food availability has increased, but the economic crisis has led to food insecurity
The challenges of food security policy in Indonesia: lessons from Vietnam, India, and Japan	A. Widiyarna, C. Wijaya, A. Wahyudi Atmoko [4]	2022	Vietnam transformed from a food crisis country to a food-secure nation. India developed institutions for food policy, while Japan invested in agricultural technology

(continued)

Table 11.1 (continued)

Paper title	Authors	Year	Main findings
Respecting, protecting, and fulfilling the right to food and nutrition during the economic crisis: the Indonesian perspective	Soekirman [23]	1999	Indonesia's food law defines food security as household access to adequate and affordable food. The Social Safety Net (SSN) program was introduced to reduce malnutrition and ensure health services
Food security management for Indonesia: the strategy during the COVID-19 pandemic	S. Darma, D. Darma [6]	2020	Indonesia ensured food security during COVID-19 through seed assistance, price stabilization, and food distribution strategies. Lessons can contribute to global food security efforts
Indonesia's regional food security in light of the impending global food crisis	N. Anwar [2]	2022	Food security in Indonesia is affected by rice production, population density, and purchasing power. Government policies should integrate land, water, and fertilizer management
Indonesian food policy insights and challenges of integrating food security and food sovereignty	Cisma Tami Voletta [26]	2016	Indonesia has a complex food policy landscape. Integrating food security with sovereignty requires better coordination, resources, and data
Sustainable integrated farming system: a solution for national food security	M. Ansar, Fadhlurrohman [3]	2018	The world food crisis threatens Indonesia. Food diversification and integrated farming systems promote sustainable agriculture
National food-crop policy model for Indonesia	P. Heytens, W. Meyers [12]	1990	Indonesia's food policy aimed at rice self-sufficiency through investment and price support. After 1985, the focus shifted to secondary crops
Food security policies in Maritime Southeast Asia	Zamroni Salim [28]	2010	The 2007/08 global food crisis affected Indonesia. Food insecurity is due to lack of access rather than supply shortages. Policies focusing on self-sufficiency have not improved food accessibility

11.3 Research Method

This study utilizes a mixed-method methodology, integrating both quantitative and qualitative techniques to thoroughly examine the Global Food Security Index (GFSI) in the specific context of Indonesia. The research utilizes both deductive and inductive methodologies. The deductive approach encompasses the examination and evaluation of preexisting theories and hypotheses pertaining to the domains of food security and the Global Food Security Index (GFSI). The utilization of the inductive technique facilitates the exploration and identification of novel insights by means of qualitative data analysis.

11.3.1 Data Source

The main data source utilized in this study comprises a comprehensive assortment of preexisting studies and statistics pertaining to the issue of food security in Indonesia. The data for the Global Food Security Index (GFSI) is sourced from reputable entities such as the Economist Intelligence Unit (EIU), the World Food Programme (WFP), and the Food and Agriculture Organization (FAO). Furthermore, the collection of qualitative data in the Indonesian food security sector involves the use of interviews and surveys administered to important players. The collection of quantitative data is facilitated by the study of the Global Food Security Index (GFSI) scores and other pertinent indicators. Qualitative data is obtained by conducting semi-structured interviews with various stakeholders, including government officials, food security specialists, and representatives from nongovernmental organizations. In order to assess views of food security, surveys are also conducted among a subset of the population. The utilization of a research technique that integrates both quantitative and qualitative data will enable a thorough investigation of the Global Food Security Index within the specific context of Indonesia. This approach will provide valuable insights into the obstacles and potential avenues for enhancing food security in the region.

11.3.2 Keywords and Filters

The essential terms for conducting data retrieval and compilation encompass the concepts of "food security," "Global Food Security Index," "Indonesia," "hunger," and "sustainable agriculture." In order to ascertain the pertinence and dependability of the sources, filters are employed, with a specific emphasis on recent publications subsequent to the year 2010 and scholarly research that has undergone peer review.

TITLE-ABS-KEY ("food security" OR "food security index") AND PUBYEAR > 2011 AND PUBYEAR < 2023 AND (LIMIT-TO (DOCTYPE, "ar")

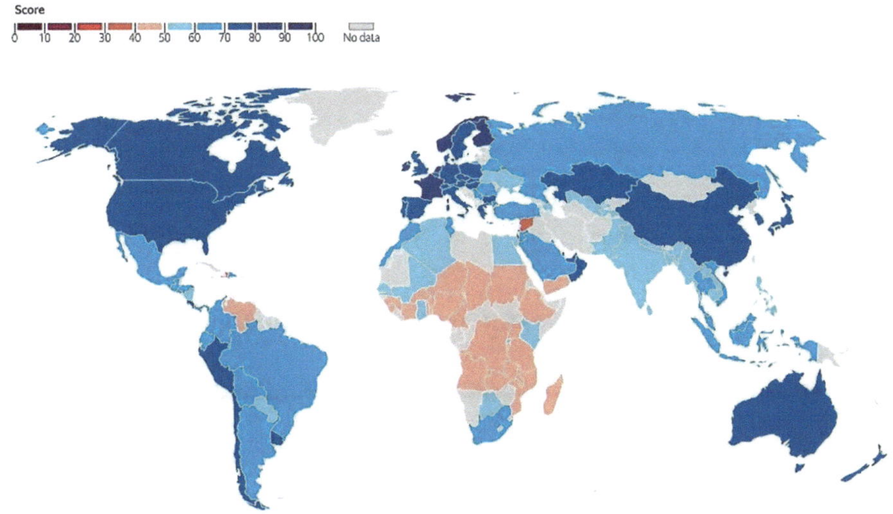

Fig. 11.3 Global Food Security Index score filled map—Source EIU

OR LIMIT-TO (DOCTYPE, "cp")) AND (LIMIT-TO (EXACTKEYWORD, "Food Security") OR LIMIT-TO (EXACTKEYWORD, "Food Supply") OR LIMIT-TO (EXACTKEYWORD, "Climate Change")) AND (LIMIT-TO (OA, "all")) AND (LIMIT-TO (LANGUAGE, "English")) AND (LIMIT-TO (PUBSTAGE, "final")) AND (LIMIT-TO (SUBJAREA, "SOCI")) AND (LIMIT-TO (EXACTSRCTITLE, "Food Policy") OR LIMIT-TO (EXACTSRCTITLE, "Food Security")) (Fig. 11.3)

11.4 Results and Discussion

Based on the extensive research and analysis conducted on this topic, the discovered outcomes will be discussed in detail in the section listed below:

11.4.1 Comparing the Global Food Security Index

11.4.1.1 Challenges and Improvements

Figure 11.4 depicts Indonesia's Global Food Security Index Scores, which were analyzed carefully. The following are some of the most important factors underlying the trend of countries' scores:

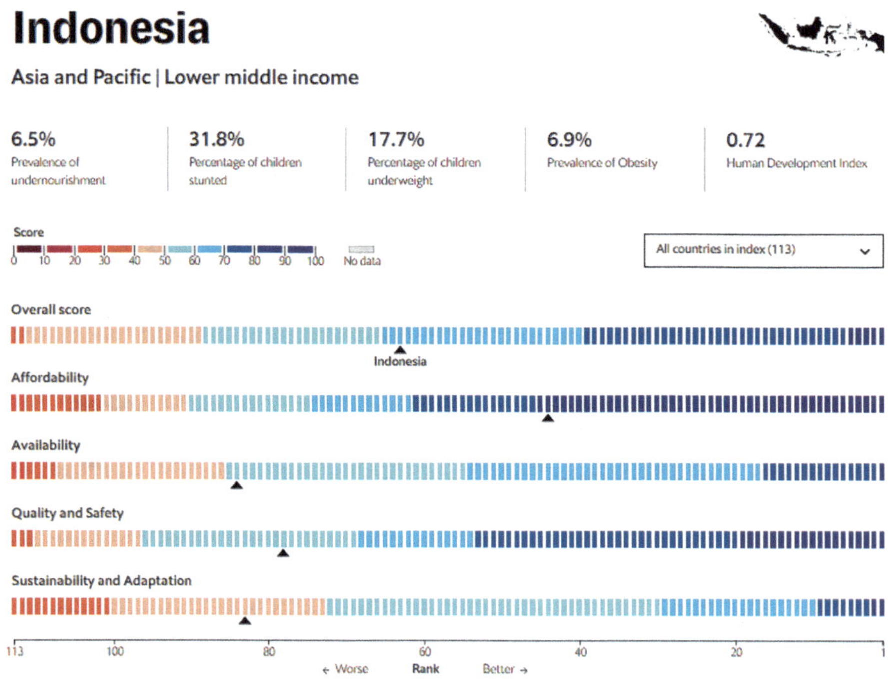

Fig. 11.4 Overall GFSI score of Indonesia

- The availability and accuracy of data play a crucial role in conducting a comprehensive assessment of Indonesia's food security indicators. However, the current situation is characterized by limited access to reliable and up-to-date data, which poses a significant challenge to obtaining a thorough understanding of Indonesia's food security status. Enhancements in data gathering and reporting procedures are necessary.
- Regional disparities: Indonesia, as a nation characterized by its diversity, exhibits notable disparities in terms of food security throughout its many areas. In order to effectively address regional discrepancies in food security, it is imperative to conduct a more detailed study that allows for targeted actions.
- The examination of socioeconomic aspects pertaining to food security, encompassing income disparity, work prospects, and educational accessibility, needs a more comprehensive analysis in order to formulate focused policy interventions for enhancement.
- The comprehensive examination of the correlation between environmental challenges, such as climate change, deforestation, natural resource management, and food security, is vital for the promotion of environmental sustainability. The implementation of sustainable practices is of paramount importance in ensuring the attainment of long-term food security.

- The assessment of food waste and loss along the supply chain, spanning from production to consumption, is crucial in determining the magnitude of this issue and pinpointing potential avenues for enhancing efficiency and augmenting food accessibility.
- The evaluation of the efficacy of governmental policies, laws, and international partnerships in advancing food security is of utmost importance in the realm of policy and governance. This analysis provides a basis for deriving recommendations aimed at policy reforms and enhanced governance.

11.4.2 Comparing Global Food Security Index Indicators

11.4.2.1 Comparing Affordability Indicator

This study assesses the capacity of consumers to procure food, their susceptibility to fluctuations in prices, and the existence of initiatives and policies aimed at assisting customers during such periods of instability.

Challenges and Improvements

Figure 11.5 depicts the following most important factors underlying the trend of country scores:

- Obtaining precise and current data pertaining to food costs, income levels, and other pertinent elements in Indonesia is a considerable challenge in terms of both availability and quality. The variability of data quality and consistency across different places and historical periods poses challenges in ensuring the comprehensiveness and reliability of the dataset.
- Indonesia, as a geographically expansive and culturally varied nation, exhibits notable regional inequalities. The consideration of regional variances is crucial

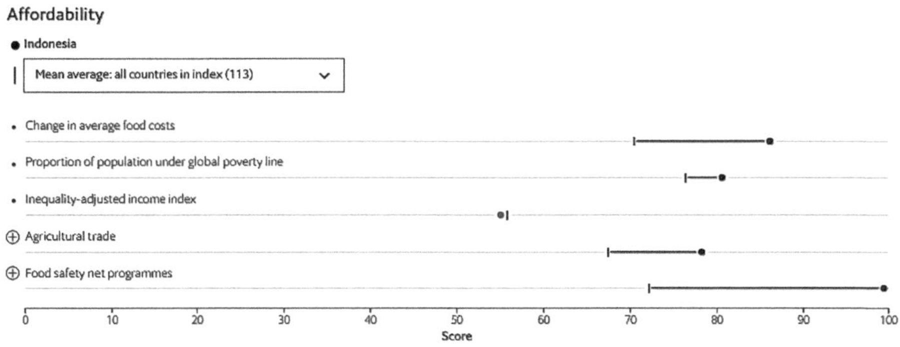

Fig. 11.5 Overall GFSI score of Indonesia

when analyzing affordability indicators since it introduces complexities in assessing food security at a national level.

- Informal food systems: A substantial segment of the Indonesian populace is dependent on informal food systems, exemplified as traditional markets. The omission of these transactions from the record poses a difficulty in including them in the calculations of affordability.
- The consumption patterns of the Indonesian population are undergoing transformation as a result of increasing income levels and shifting lifestyles. The ever-changing nature of consumer behavior makes it difficult to evaluate the affordability of food and its implications for food security.
- The affordability of food can be significantly influenced by fluctuations in inflation rates and exchange rates. The precise integration of these economic indicators into the affordability indicator is of utmost importance. However, it can pose challenges due to their inherent volatility.
- The examination of the influence of government policies and socioeconomic factors on the affordability of food is crucial for comprehensive understanding. Nevertheless, the impact of policies and cultural factors on food security is intricate and diverse, posing difficulties in accurately measuring its consequences.

11.4.2.2 Comparing Availability Indicators

This study assesses agricultural production and on-farm capabilities, evaluates the potential danger of supply disruption, examines the national capacity to spread food, and analyzes research initiatives aimed at expanding agricultural output.

Challenges and Improvements
Figure 11.6 depicts the following some of the most important factors underlying the trend of country scores:

- The data quality and availability pertaining to food security in Indonesia may exhibit inconsistencies or incompleteness, hence posing challenges in acquiring accurate and current information necessary for the computation of the index.
- Indonesia, being a geographically extensive and culturally diverse nation, exhibits notable regional inequalities in terms of food security. In order to offer a thorough and accurate evaluation, it is imperative for researchers to acknowledge and tackle these disparities.
- Indonesia encounters swings in food production and availability during different seasons, mostly influenced by weather patterns and agricultural cycles. Consequently, it becomes imperative to consider and accommodate these seasonal variations.
- The country's data collection infrastructure may provide difficulties that could impede the comprehensive gathering of food security data, hence resulting in potential gaps within the dataset.

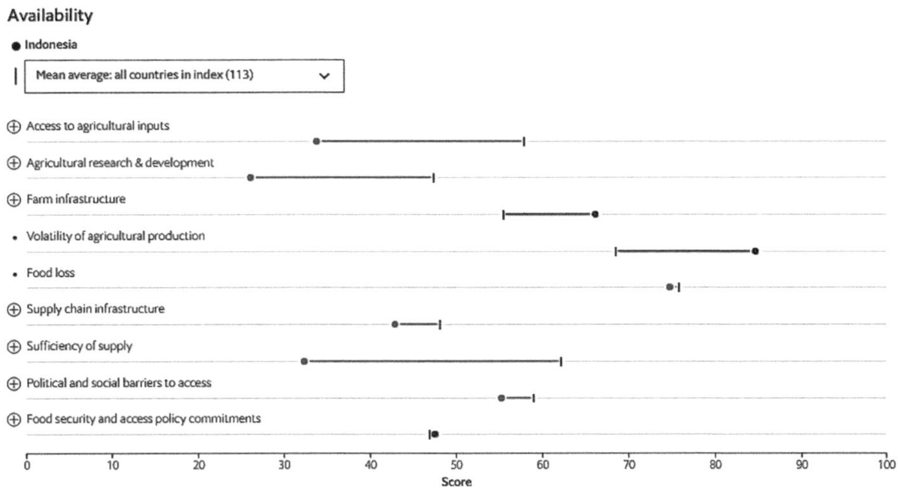

Fig. 11.6 Availability score of Indonesia

- The intricate and disjointed structure of Indonesia's food supply networks presents difficulties in evaluating food availability, particularly in rural and less accessible regions.
- Cultural and social variables play a crucial role in shaping food availability in Indonesia, as they influence local food consumption preferences. When evaluating data, it is imperative for researchers to take into account these elements.

11.4.2.3 Comparing Quality and Safety Indicators

This study assesses the diversity and nutritional value of typical dietary patterns, together with the aspect of food safety.

Challenges and Improvements
Figure 11.7 depicts the following some of the most important factors underlying the trend of country scores:

- The acquisition of precise and current data regarding food quality and safety indicators in Indonesia presents difficulties due to disparities in data collection and reporting methods across different regions and sources.
- The utilization of standardized criteria in the Global Food Security Index is apparent, yet the proper representation of Indonesia's food system's distinctive issues and peculiarities can provide a challenge.
- Indonesia has notable regional disparities in terms of food production and safety, owing to its unique geographical and cultural characteristics. It is vital to comprehend these differences and their ramifications on the total score.

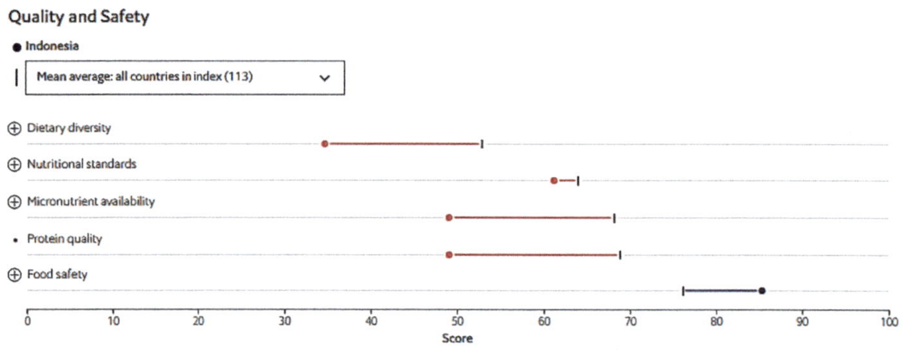

Fig. 11.7 Quality and safety score of Indonesia

- The accuracy of the indicator can be compromised by inconsistencies in the enforcement of food safety rules. It is imperative to consider these variations in the analysis.
- The influence of cultural practices and consumer behaviors in Indonesia can have implications for food safety. It is imperative to investigate these aspects and their impact on the indication in order to conduct a thorough evaluation.
- The susceptibility of Indonesia to climate change has the potential to impact its food security. It is of utmost importance to modify the indicator in order to incorporate considerations for climate-related difficulties, as this is essential for achieving a comprehensive assessment.

11.4.2.4 Comparing Sustainability and Adaptation Indicators

This study evaluates the level of vulnerability of a nation to the consequences of climate change, its proneness to hazards associated with natural resources, and the extent to which the country is implementing measures to adapt to these risks.

Challenges and Improvements

Figure 11.8 depicts following some of the most important factors underlying the trend of country scores:

- Indonesia, as a geographically expansive and culturally diverse nation, exhibits notable regional disparities in relation to climatic patterns, agricultural practices, and socioeconomic circumstances. The presence of these variances poses a challenge in developing a universally applicable set of sustainability and adaptation indicators that effectively capture the food security status of a given nation.
- Cultural and social issues are of paramount importance in the context of food security. Incorporating local customs, food preferences, and traditional agricultural practices into a standardized index is a hard and time-consuming task that is crucial for achieving a full assessment.

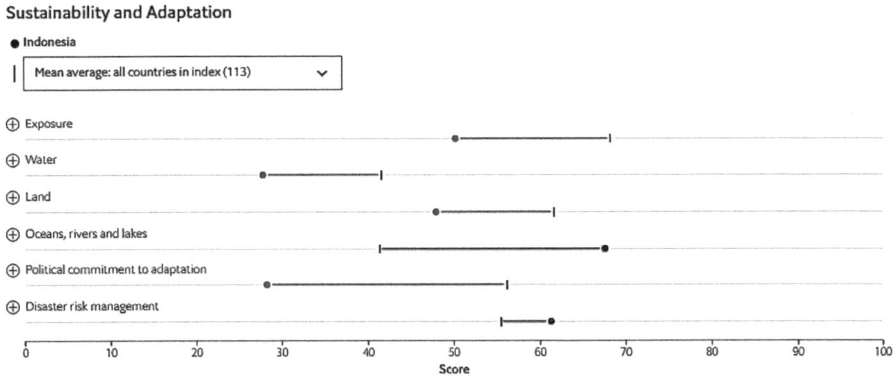

Fig. 11.8 Sustainability and adaptation score of Indonesia

- Customizing indicators is a valuable approach to enhance the accuracy of assessing food security by tailoring sustainability and adaptation indicators to individual regions or provinces within Indonesia. This approach acknowledges the heterogeneous conditions and requirements of many regions within the nation.
- The implementation of a system for continuous data monitoring and updates can effectively guarantee the ongoing relevance and currency of the indicators. The establishment and maintenance of a real-time database necessitates collaborative efforts with local authorities and organizations.
- Multi-stakeholder engagement is a valuable approach that can contribute to the improvement of indicator quality and reliability. This approach involves the active participation of several stakeholders, such as government agencies, nongovernmental organizations, academics, and local communities. By involving these diverse groups, the indicators can benefit from a broader range of perspectives and expertise. Collaboration facilitates the development of a more thorough comprehension of the challenges and solutions pertaining to food security in Indonesia.
- The integration of qualitative data is crucial for gaining a comprehensive understanding of the sociocultural factors that influence food security, in addition to quantitative data. This can be achieved through the utilization of surveys, interviews, and field research methods. The integration of quantitative and qualitative methodologies has the potential to generate more comprehensive and reliable indicators.

11.4.2.5 Recommendations and Suggestions for the Future

Online E-Commerce Portal for Farmers

In order to tackle the food security situation in Indonesia, it is imperative to take a significant stride by establishing an online digital portal that facilitates direct connection between farmers and consumers. This platform has the potential to address multiple difficulties that are currently being encountered. First and foremost, the platform should incorporate functionalities that allow farmers to exhibit their agricultural products, provide real-time communication of stock availability, and establish equitable pricing. This not only confers agency to farmers but also guarantees a consistent revenue stream, thereby motivating them to increase their production. In addition, it is imperative for the platform to enable transparent and traceable supply chains, thereby instilling customers with assurance regarding the provenance and caliber of the food they acquire. Moreover, the portal has the capacity to offer instructional materials and assistance pertaining to sustainable agricultural techniques, hence enhancing crop productivity and optimizing food production efficiency. Incorporating mobile accessibility is of critical importance in order to effectively reach distant regions of Indonesia that may experience low connectivity. Finally, it is imperative to engage in partnerships with local government entities and nongovernmental organizations (NGOs) in order to guarantee adherence to regulatory standards. Additionally, there is a need for substantial investment in digital infrastructure to provide universal accessibility to this portal. Through the implementation of these proposals, the online digital portal has the potential to significantly contribute to the mitigation of the food security issue in Indonesia. This can be achieved by facilitating direct connections between farmers and consumers, supporting equitable trading practices, and encouraging the use of sustainable agricultural methods (Fig. 11.9).

Targeted Digital Food Subsidy

Individuals utilized their national identity card or government-provided food card to procure food commodities from establishments that offer subsidies depending on their income or cater to those who are susceptible to food crises. The proposed academic revision: The proposed study aims to examine the connection and integration of various retail establishments such as malls, marts, utility stores, and small food stores. Additionally, it seeks to explore the implementation of a digital inventory system that is integrated with the government's food subsidy central system. In the event that an entity lacks a digital inventory system, the government will furnish them with a mobile point-of-sale (POS) system, which can be utilized to establish a connection with their inventory. The determination of eligibility for food subsidy is facilitated through the government's integrated system, wherein individuals' personal identity numbers are used as a basis for purchasing items from various shops. Once eligibility is confirmed through the banking system, a real-time settlement of the subsidy amount and incentive to shop is made from the government's end. The Digital Food Subsidy Programme enables the government to determine the eligibility criteria for individuals who will receive subsidies, specify

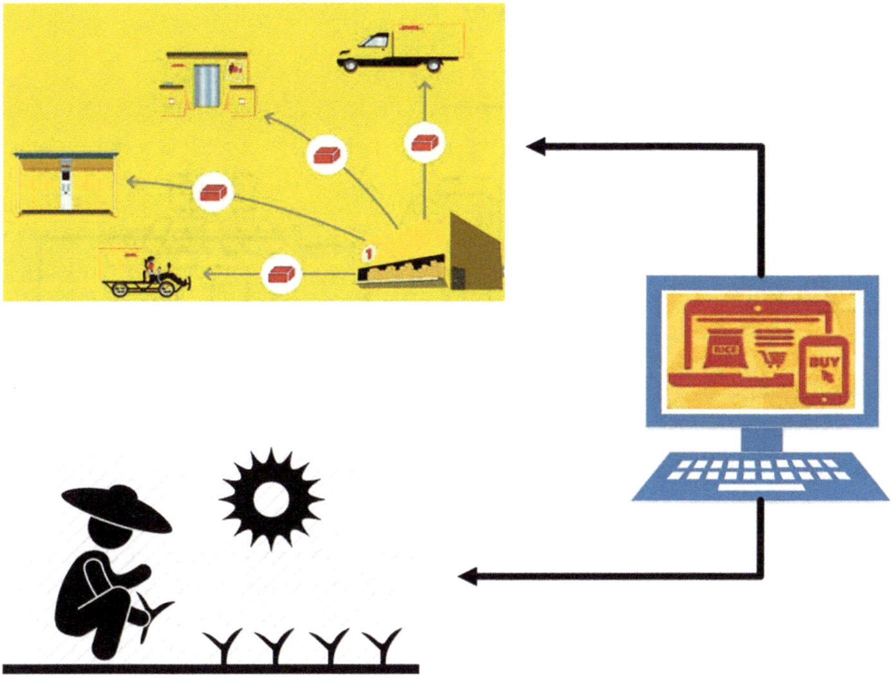

Fig. 11.9 Working of this portal

the percentage of subsidy to be allocated, and designate the specific food products eligible for subsidy. Additionally, the government has the potential to broaden its scope of coverage beyond food products to include pharmacy services, fuel stations, and other relevant sectors in order to offer specific digital subsidies (Fig. 11.10).

11.4.2.6 Free Chicken Policy

The government would allocate chicks to individuals facing socioeconomic disadvantages, thereafter providing them with training on utilizing these resources to mitigate food crises. A chicken can be seen as a financial instrument that serves as a means of tied investment or forced saving, thereby discouraging individuals from engaging in discretionary expenditures. Chickens are characterized by their relatively modest capital requirements and rapid turnover rates. Individuals commonly utilize eggs and meat derived from chickens as a primary source of sustenance while also having the potential to generate income through their sales. The implementation of the chicken intervention necessitates the establishment of complementary infrastructure, as it cannot be executed in isolation. It is imperative to establish supplementary amenities such as veterinary services and training support. Furthermore, it is imperative to undertake substantial efforts in order

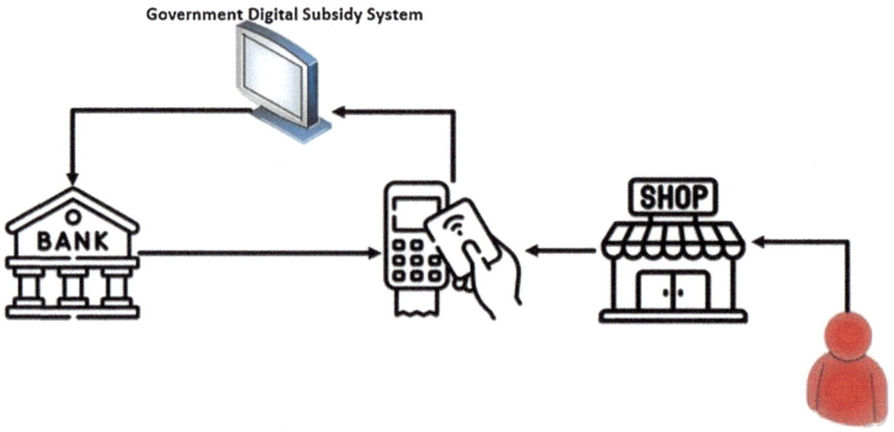

Fig. 11.10 Working of this system

to facilitate individuals' access to marketplaces and enhance market connections for the purpose of selling eggs and chickens. In addition, it is important to establish a series of measures aimed at effectively coordinating individuals to leverage economies of scale and establish connections with markets collaboratively (Fig. 11.11).

11.4.2.7 Policy to Reduce Food Wastage

Government programs play a pivotal role in educating individuals toward more sustainable practices in order to address the food security challenge and mitigate food wastage in Indonesia. Primarily, it is imperative for the government to initiate educational initiatives aimed at enhancing public knowledge of the need to reduce food waste. Efforts to promote efficient meal planning and good food storage can encompass several strategies, such as the dissemination of public service announcements, implementation of school programs, and organization of community workshops. These initiatives aim to educate individuals and households on the principles and techniques of effective meal planning and appropriate food preservation methods. Additionally, implementing incentives for the food business to provide surplus or unsold food to charitable organizations and food banks has the potential to substantially mitigate waste while simultaneously addressing the needs of disadvantaged individuals. Furthermore, it is imperative for the government to actively promote the advancement of sustainable agricultural practices and the implementation of effective supply chain management in order to mitigate food loss during both the production and distribution stages. In addition, the adoption of policies aimed at fostering the utilization of technology and innovation in the realm of food preservation and distribution can effectively contribute to the

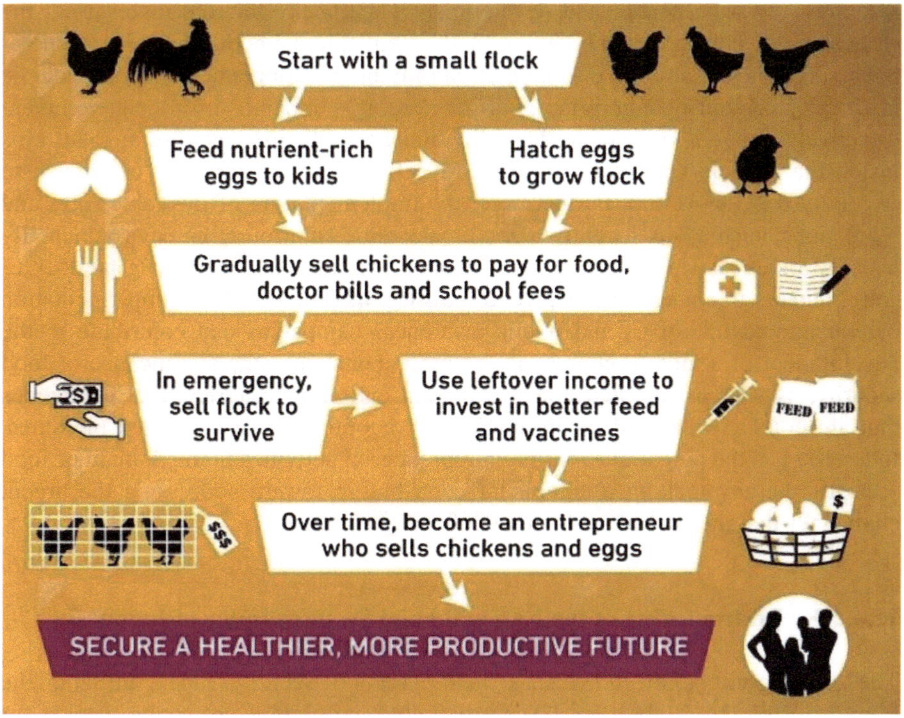

Fig. 11.11 Working of this policy

elongation of perishable goods' lifespan. In conclusion, the endorsement of local, small-scale farming endeavors and the advocacy for the consumption of locally procured, seasonal food items can contribute to the advancement of food security through a diminished reliance on imported goods. The integration of comprehensive government efforts, alongside active community engagement and collaboration with the commercial sector, holds the potential to make substantial contributions toward the reduction of food loss and the enhancement of food security in Indonesia.

11.4.2.8 Promoting Start-ups Transforming Food Security Problem

In order to effectively mitigate the food security crisis in Indonesia and foster the growth of start-ups with the potential to make substantial contributions in arid regions, it is advisable to contemplate a range of proposals and ideas. Initially, it is imperative for the government to construct a comprehensive support structure that specifically caters to start-up enterprises, with a particular emphasis on developing new solutions to address the issue of food security in arid regions. This includes the provision of financial incentives, grants, and improved access to finance, coupled with the streamlining of bureaucratic processes. Furthermore, the allocation of resources toward research and development plays a pivotal role in fostering the emergence of innovative technologies and sustainable methodologies. It is imperative to promote collaborative collaborations among governmental entities, academic institutes, and start-up enterprises. Furthermore, it is imperative for the government to actively support knowledge dissemination and establish networking platforms inside the start-up ecosystem in order to cultivate an environment conducive to creativity and innovation. Furthermore, the implementation of favorable rules and regulations, such as tax incentives and breaks aimed at promoting environmentally friendly practices, might serve as a catalyst for attracting a greater number of entrepreneurs to engage in this pivotal industry. In conclusion, the implementation of educational initiatives and public awareness campaigns can contribute to the development of a more comprehensive comprehension of the significance of food security, as well as facilitate the widespread adoption of agricultural techniques that are conducive to arid environments. By adopting these suggested measures, Indonesia has the potential to foster entrepreneurial activities in the domain of food security, thereby making a significant contribution toward addressing the urgent challenges that the country currently confronts in this area.

11.4.2.9 Restructuring of the National Food Security Index of Indonesia

The reconfiguration of Indonesia's National Food Security Index, utilizing the framework of the Global Food Security Index, presents a promising avenue for improving the nation's strategy in tackling its food security issues. The initial step for the Indonesian government should involve the incorporation of essential indicators and metrics employed in the Global Food Security Index. These indicators encompass crucial elements such as the cost, availability, quality, and safety of food. This approach will yield a more thorough evaluation of the country's food security situation. In addition, it is imperative to underscore the significance of sustainable agriculture and food production practices in order to guarantee enduring resilience. It is advisable to provide recommendations that encompass several tactics aimed at mitigating post-harvest losses, fostering crop diversification, and offering assistance to small-scale farmers. Furthermore, endeavors aimed at enhancing infrastructure, including transportation and storage facilities, can contribute to the optimization of food distribution efficiency. Enhancing social safety nets and reducing economic inequality are crucial elements in resolving food security concerns since they have a significant role in improving the accessibility of food for communities that are particularly vulnerable. Furthermore, allocating resources toward research and development in the field of innovative farming technology has the potential to augment productivity. The implementation of measures aimed at mitigating climate change and its repercussions on agriculture is of utmost importance. This is due to the fact that the altering climate presents a substantial risk to Indonesia's food security. By realigning the National Food Security Index with international benchmarks and following the suggested measures, Indonesia can make substantial progress in addressing its food security challenges and establishing a more sustainable and reliable food provision for its population.

11.4.2.10 Stopping Smuggling of Food Products

In order to effectively mitigate the Indonesian food security crisis and counteract the illicit trade of food goods, a range of recommendations and ideas might be taken into account. To begin with, it is imperative to prioritize the enhancement of border security and customs control measures in order to mitigate the unlawful ingress of food goods effectively. The implementation of sophisticated technologies for the surveillance and examination of commodities at border checkpoints and harbors is encompassed in this endeavor. One potentially useful approach involves fostering collaboration with neighboring nations to facilitate the coordination of activities and exchange intelligence pertaining to smuggling networks. Furthermore, the implementation of more stringent sanctions and enhanced enforcement efforts targeting smugglers can serve as an effective deterrent. Promoting domestic food production is of utmost importance through the provision of enhanced agricultural practices, access to contemporary technology, and financial incentives to augment production, thereby assisting farmers. Promoting sustainable agricultural practices and the cultivation of diverse crops has the potential to mitigate reliance on certain food commodities that are susceptible to smuggling. In addition, the allocation of resources toward the development of transport and distribution infrastructure can play a crucial role in facilitating the effective delivery of food to places that are now lacking adequate access. This, in turn, can contribute to a decrease in the motivation for engaging in illicit smuggling activities. Ultimately, the dissemination of information to the general public regarding the adverse consequences of food smuggling on both food security and the economy has the potential to generate backing for endeavors aimed at combating smuggling activities. It is imperative to adopt a comprehensive approach that integrates several tactics in order to effectively address the food security situation in Indonesia and mitigate the illicit trade of food goods.

11.4.2.11 Promote Domestically Created Food Products

Advocating for the prioritization of domestically produced food items above imported alternatives emerges as a pivotal proposition in tackling the prevailing food security predicament in Indonesia. In order to address this matter in a comprehensive manner, it is important to take into account a variety of solutions. To begin with, it is imperative for the Indonesian government to allocate resources toward the development of agricultural infrastructure, technology, and research. This strategic focus will serve to augment the efficiency and excellence of domestic food production. This encompasses enhancing irrigation infrastructure, offering financial assistance to farmers, and promoting the implementation of sustainable farming techniques.

Moreover, it is imperative to implement a coordinated endeavor aimed at providing comprehensive education and training to farmers in contemporary agricultural methodologies with the objective of optimizing crop productivity. Moreover, it is imperative to implement rigorous quality control protocols in order to guarantee the safety and nutritional integrity of domestically sourced food products. The promotion of domestic products can also encompass measures such as the reduction of import taxes and the implementation of trade policies that provide incentives for the use of locally cultivated and processed foods. Finally, the utilization of public awareness campaigns can be implemented as a strategy to enlighten customers regarding the advantages associated with endorsing local agriculture, alongside emphasizing the significance of diminishing dependence on imported goods. By prioritizing these suggestions, Indonesia has the potential to enhance its food security and enhance the economic well-being of local farmers while simultaneously lowering its reliance on imported food commodities.

11.5 Conclusion

This study presents a thorough analysis of the Global Food Security Index (GFSI) in the specific context of Indonesia, revealing a complex array of obstacles and prospects that influence the country's food security. The Global Food Security Index (GFSI) has proven to be an invaluable resource, providing unique insights into crucial elements of the nation's food system. Its findings have the potential to inform and advise policymakers, stakeholders, and researchers in their endeavors to improve food security in Indonesia. The study conducted indicates that Indonesia is confronted with a complex interaction of several factors that exert an influence on food security. Although there have been advancements in some significant domains, such as the availability and accessibility of food, the nation continues to face ongoing challenges pertaining to the quality and safety of food, environmental sustainability, and nutritional consequences. The susceptibility of small-scale farmers to the impacts of climate change and the systemic obstacles in the delivery of food have become significant issues, emphasizing the necessity for comprehensive policy responses. The necessity of adopting a comprehensive strategy that encompasses the entirety of Indonesia's food value chain becomes apparent when considering the imperative of enhancing the nation's food security. The implementation of sustainable farming practices, the use of innovative technologies, and the empowerment of small-scale farmers are crucial measures in ensuring the resilience and inclusivity of the food system. Furthermore, it is imperative to prioritize the enhancement of food safety standards and the guarantee of equitable access to safe and nourishing food across all demographic groups. The importance of collaboration between government agencies, nongovernmental organizations, and international partners cannot be overstated when it comes to tackling the complex and varied issue of food security. In summary, this research highlights the significant importance of the Global Food Security Index (GFSI) in both evaluating food security and facilitating evidence-based policy transformations. In order to ensure a more prosperous and food-secure future for Indonesia, it is crucial to maintain active involvement with the Global Food Security Index (GFSI) and utilize its valuable insights as a framework to educate the country's policy-making process. This approach will pave the way toward a future that is both food secure and sustainable, benefiting the entire population of Indonesia.

Funding
The authors of the paper reported that there is no funding associated with the work featured in the article.

Conflicts of Interest
The authors declare no conflict of interest.

Acknowledgments The research work of this article was supported by Achmad Nurmandi from the Universitas Muhammadiyah Yogyakarta. The researcher would like to express his immense gratitude to his supporters, who have provided all the necessary insight and expertise to assist in this research.

References

1. Ajaero, C. K. (2017). A gender perspective on the impact of flood on the food security of households in rural communities of Anambra state, Nigeria. *Food Security, 9*(4), 685–695. https://doi.org/10.1007/s12571-017-0695-x
2. Anwar, N. (2022). INDONESIA'S REGIONAL FOOD SECURITY IN LIGHT OF THE IMPENDING GLOBAL FOOD CRISIS. TRIKONOMIKA: Jurnal Ekonomi, 21(2), 101–110. https://doi.org/10.23969/trikonomika.v21i2.7113
3. Ansar, M. and Fathurrahman F. (2018). Sustainable integrated farming system: A solution for national food security and sovereignty. IOP Conference Series Earth and Environmental Science 157(1):012061. https://doi.org/10.1088/1755-1315/157/1/012061
4. Atmoko, A. W., & Wijaya, C. (2022). The Challenges of Food Security Policy in Indonesia: Lesson Learned from Vietnam, India, and Japan. Technium Social Sciences Journal, 33, 1–15. https://doi.org/10.47577/tssj.v33i1.6937
5. Candel, J. J. L., & Biesbroek, R. (2018). Policy integration in the EU governance of global food security. *Food Security, 10*(1), 195–209. https://doi.org/10.1007/s12571-017-0752-5
6. Darma, Surya, and Dio Caisar Darma. (2020) Food Security Management for Indonesia: The Strategy during the Covid-19 Pandemic. Management Dynamics in the Knowledge Economy, vol. 8, no. 4, Sciendo, 2020, pp. 371–381. https://doi.org/10.2478/mdke-2020-0024
7. Das, M., Sharma, A., & Babu, S. C. (2018). Pathways from agriculture-to-nutrition in India: implications for sustainable development goals. *Food Security, 10*(6), 1561–1576. https://doi.org/10.1007/s12571-018-0858-4
8. Devereux, S., Béné, C., & Hoddinott, J. (2020). Conceptualizing COVID-19's impacts on household food security. *Food Security*, *12*(4), 769–772. https://doi.org/10.1007/s12571-020-01085-0
9. Dillon, H. S. (2004) Food Security in Indonesia, Agricultural and Food Sciences, Economics. https://api.semanticscholar.org/CorpusID:154732344
10. Fiorella, K. J., Hickey, M. D., Salmen, C. R., Nagata, J. M., Mattah, B., Magerenge, R., Cohen, C. R., Bukusi, E. A., Brashares, J. S., & Fernald, L. H. (2014). Fishing for food?

Analyzing links between fishing livelihoods and food security around Lake Victoria, Kenya. *Food Security*, *6*(6), 851–860. https://doi.org/10.1007/s12571-014-0393-x
11. Haysom, G., & Tawodzera, G. (2018). “Measurement drives diagnosis and response”: Gaps in transferring food security assessment to the urban scale. *Food Policy*, *74*, 117–125. https://doi.org/10.1016/j.foodpol.2017.12.001
12. Heytens, P. M. & Meyers W. H. (1990). National Food-Crop Policy Model for Indonesia. Center for Agricultural and Rural Development (CARD) Publications 90-tr16, Center for Agricultural and Rural Development (CARD) at Iowa State University.
13. Hellegers, P. (2022). Food security vulnerability due to trade dependencies on Russia and Ukraine. *Food Security*, *14*(6), 1503–1510. https://doi.org/10.1007/s12571-022-01306-8
14. Izraelov, M., & Silber, J. (2019). An assessment of the global food security index. *Food Security, 11*(5), 1135–1152. https://doi.org/10.1007/s12571-019-00941-y
15. Lancker, K., Fricke, L., & Schmidt, J. O. (2019). Assessing the contribution of artisanal fisheries to food security: A bio-economic modeling approach. *Food Policy, 87*. https://doi.org/10.1016/j.foodpol.2019.101740
16. Limenta, Michelle Engel and Chandra, Sianti (2017) INDONESIAN FOOD SECURITY POLICY,” Indonesia. Law Review: Vol. 7 : No. 2, Article 6. https://doi.org/10.15742/ilrev.v7n2.198
17. Mekonnen, D. A., & Gerber, N. (2017). Aspirations and food security in rural Ethiopia. *Food Security, 9*(2), 371–385. https://doi.org/10.1007/s12571-017-0654-6
18. Nawrotzki, R. J., Robson, K., Gutilla, M. J., Hunter, L. M., Twine, W., & Norlund, P. (2014). Exploring the impact of the 2008 global food crisis on food security among vulnerable households in rural South Africa. *Food Security, 6*(2), 283–297. https://doi.org/10.1007/s12571-014-0336-6
19. Papworth, A., Maslin, M., & Randalls, S. (2022). The challenges of a food sovereignty perspective: an analysis of the foodways of the Rama indigenous group, Nicaragua. *Food Security, 14*(4), 1013–1026. https://doi.org/10.1007/s12571-022-01268-x
20. Piesse, M. (2016). Food security in Indonesia: A continued reliance on foreign markets. Independent Strategic Analysis of Australia Global Interests.
21. Rafai, I. (2014). The law no. 18/2012 governing food security in Indonesia
22. Randell, H., Gray, C., & Shayo, E. H. (2022). Climatic conditions and household food security: Evidence from Tanzania. *Food Policy, 112*. https://doi.org/10.1016/j.foodpol.2022.102362
23. Soekirman. (1999) Respecting, protecting and fulfilling the right to food and nutrition during the economic crisis: the Indonesian perspective. SCN News. 1999 Jul;(18):56–9. PMID: 12290440.
24. Thow, A. M., Greenberg, S., Hara, M., Friel, S., duToit, A., & Sanders, D. (2018). Improving policy coherence for food security and nutrition in South Africa: a qualitative policy analysis. *Food Security, 10*(4), 1105–1130. https://doi.org/10.1007/s12571-018-0813-4
25. Vaitla, B., Coates, J., Glaeser, L., Hillbruner, C., Biswal, P., & Maxwell, D. (2017). The measurement of household food security: Correlation and latent variable analysis of alternative indicators in a large multi-country dataset. *Food Policy, 68*, 193–205. https://doi.org/10.1016/j.foodpol.2017.02.006
26. Voletta, Cisma Tami. (2016). Indonesian food policy insights and challenges of integrating food security and food sovereignty. Agrarian, Food and Environmental Studies (AFES). Retrieved from http://hdl.handle.net/2105/37256
27. Weikard, H.-P. (2016). Phosphorus recycling and food security in the long run: a conceptual modeling approach. *Food Security, 8*(2), 405–414. https://doi.org/10.1007/s12571-016-0551-4
28. Zamroni, S. (2010). Food security policies in maritime Southeast Asia. Canadian Electronic Library.

GPSR Compliance

The European Union's (EU) General Product Safety Regulation (GPSR) is a set of rules that requires consumer products to be safe and our obligations to ensure this.

If you have any concerns about our products, you can contact us on ProductSafety@springernature.com

In case Publisher is established outside the EU, the EU authorized representative is:

Springer Nature Customer Service Center GmbH
Europaplatz 3
69115 Heidelberg, Germany

Batch number: 10725825

Printed by Printforce, the Netherlands